"This book is a welcome addition to the literature on economic dynamics. Its clear writing style and the emphasis on coding using MATLAB® make it a compelling text for introducing undergraduate economics students to stability issues, cycles, and growth. The emphasis on both standard models like the Solow growth model and less standard ones such as the Goodwin growth cycle appeal to a broad spectrum of economists in the profession, and the highly competent authors have put years of experience with the material into a highly accessible textbook. I highly recommend the book to anyone willing to incorporate numerical methods into macroeconomic courses."

Daniele Tavani, *Associate Professor, Colorado State University*

AN INTRODUCTION TO ECONOMIC DYNAMICS

An Introduction to Economic Dynamics provides a framework for students to appreciate and understand the basic intuition behind economic models and to experiment with those models using simulation techniques in MATLAB®.

This book goes beyond the often-limited scope of other texts on economic models, which have largely focused on elucidating static equilibrium models. Comparative static analysis inhibits students from asking how the equilibrium position is achieved from an initial out-of-equilibrium position and limits their understanding of the dynamics that underlie such analysis. In this textbook, readers are introduced to ten well-established macroeconomic models – including Keynesian multiplier models, Samuelson's multiplier and Solow's growth model – and guided through the dynamical systems behind each model. Every chapter begins with an overview of the economic problem which the model is designed to help solve followed by an explanation of the mathematics of the model. Solutions are provided using simulation and visualisation techniques in MATLAB®, which are interwoven organically with the analysis and are introduced in a step-by-step fashion to guide the reader along the way. Appendices provide an introduction to MATLAB® along with all the necessary codes.

The book is ideally suited for courses in economic dynamics, macroeconomic modelling and computational economics, as well as for students of finance, mathematics and engineering who are interested in economic models.

Srinivas Raghavendra is an Associate Professor in Economics at the Azim Premji University, Bangalore, India and with the Department of Economics at the University of Galway in Ireland.

Petri T. Piiroinen is an Associate Professor in Nonlinear Mechanics at Chalmers University of Technology, Göteborg, Sweden.

Routledge Advanced Texts in Economics and Finance

32 The Economics of Transition
Developing and Reforming Emerging Economies
Edited by Ichiro Iwasaki

33 Applied Spatial Statistics and Econometrics
Data Analysis in R
Edited by Katarzyna Kopczewska

34 Spatial Microeconometrics
Giuseppe Arbia, Giuseppe Espa and Diego Giuliani

35 Financial Risk Management and Derivative Instruments
Michael Dempsey

36 The Essentials of Machine Learning in Finance and Accounting
Edited by Mohammad Zoynul Abedin, M. Kabir Hassan, Petr Hajek and Mohammed Mohi Uddin

37 Financial Economics and Econometrics
Nikiforos T. Laopodis

38 Applied Welfare Economics, Second Edition
Cost-Benefit Analysis of Projects and Policies
Massimo Florio and Chiara Pancotti

39 An Introduction to Economic Dynamics
Modelling, Analysis and Simulation
Srinivas Raghavendra and Petri T. Piiroinen

For more information about this series, please visit: www.routledge.com/Routledge-Advanced-Texts-in-Economics-and-Finance/book-series/SE0757

AN INTRODUCTION TO ECONOMIC DYNAMICS

Modelling, Analysis and Simulation

SRINIVAS RAGHAVENDRA AND PETRI T. PIIROINEN

LONDON AND NEW YORK

Designed cover image: © Getty Images / Sono Creative

First published 2023
by Routledge
4 Park Square, Milton Park, Abingdon, Oxon OX14 4RN

and by Routledge
605 Third Avenue, New York, NY 10158

Routledge is an imprint of the Taylor & Francis Group, an informa business

For MATLAB and Simulink product information, please contact:
The MathWorks, Inc. 3 Apple Hill Drive Natick, MA, 01760-2098 USA Tel: 508-647-7000 Fax: 508-647-7001 E-mail: info@mathworks.com

British Library Cataloguing-in-Publication Data
A catalogue record for this book is available from the British Library

ISBN: 978-0-367-34190-9 (hbk)
ISBN: 978-0-367-34189-3 (pbk)
ISBN: 978-0-429-32440-6 (ebk)

DOI: 10.4324/9780429324406

Typeset in Times New Roman
by Apex CoVantage, LLC

To our parents

CONTENTS

PREFACE

This book aims to introduce *economic dynamics* to the undergraduate students of economics. Undergraduate economics education remains largely focused on elucidating static equilibrium models. The ensuing comparative static analysis inhibits students from asking how an equilibrium is reached from an initial out-of-equilibrium position and limits their understanding of the dynamics that underlie such analysis. In turn, static models and methods have become a standard mode of thinking about real economic problems and the associated analytical tools and techniques are used for exploring economic reality, i.e. data, in the wider policy analysis. This self-reinforcing dynamic has locked-in much of the undergraduate economic teaching in the static mode and has also limited the scope of the undergraduate textbook market. The aim of this book is to contribute to this challenge in a modest way by introducing economic dynamics to undergraduate economics students.

The idea for the book arose from our long-standing collaboration in teaching a joint module called *Modelling, Analysis and Simulation*, which was offered to the second-year B.Sc. program in *Financial Mathematics and Economics* at the National University of Ireland Galway (NUI Galway) in Ireland. The B.Sc. program, which is one of the most successful undergraduate programs, is an interdisciplinary program taught between the School of Mathematics, Statistics and Applied Mathematics and the School of Business and Economics.

Around 2009, following an external review of the B.Sc. program, there was an opportunity to introduce a new module. At that time we were working together on a few research projects in economics, particularly looking at business cycle models. From our own discussions on various aspects of economics, Petri, who is an applied mathematician specialised in non-smooth dynamics, was bemused by the lack of dynamics in economic models. Raghav, an economist trained in macroeconomics and political economy, was always aware of limitations of static equilibrium models in teaching and their usefulness for applied policy. When the request for a new module came about, we both thought it would be a great opportunity to put some of our own ideas into action and suggested that we will offer a module on economic dynamics.

Our diverse backgrounds, experiences and perspectives of teaching and learning shaped our thoughts about the content and the approach for the module. We were also cognizant of the wider context that underpinned the dominant perception and prescriptions about teaching and learning practices in the Universities, and also the experiences and expectations of the students. All things considered, we decided to design a module that would help widen students' skill set in terms of thinking about problems with an analytical and critical bent of mind, and also enhance their basic computing skills, necessary for experimentation and self-learning.

These principles informed and shaped our ideas for the new module, which we titled *Modelling, Analysis & Simulation*. In order to ease them into the dynamic way of thinking, we used some of the popular economic models that undergraduate students would have been exposed to in their first and second years of study. We used models such as the Cobweb model, the IS/LM model, the Solow-Swan model, etc., which are usually taught in the first two years of the undergraduate economics curriculum. Since students knew the comparative static analysis of such models, it was easier for us to introduce the dynamics that underpins these models. So the choice of models in the module is driven by this consideration of students' familiarity with these models rather than our personal preferences. However, familiarity is not the only requirement for this book and it is written in such a way that it would be accessible for students from other disciplines who are not familiar with these models.

In terms of the approach, contrary to the usual mathematics-led approach, we use a more inclusive organic approach of building the models from the bottom up. The chapters start with the economic logic and central questions of the model. We then discuss how they are translated into mathematical models, and then help the reader understand the analysis of the models, both qualitatively using simulation, and quantitatively through analytical methods. We felt that such a "bottom-up" approach would be more inclusive since students would always be guided by the economic logic that underpins the (scary!) mathematical equations and learn them in an intuitive manner. We also felt that the aspect of experimentation in learning economics would empower students as they would know exactly how the model is set up from scratch and that they can play with the model using simulation. The latter aspect provides the avenue for self-learning and we have often seen that students feel more confident and sure-footed in their understanding of economic models when they "discover" various unstated properties of the model on their own.

Furthermore, experimentation using MATLAB® would enable students to learn more about programming and allow them to work with computer codes more freely, as opposed to the menu-driven computational software like spreadsheet-based programs. Moreover, the approach used in the book should help students learn MATLAB® coding in the context of economic models and such a context-driven approach will help students to acquire the programming and simulation skills in a more assured way. We hope that the MATLAB® codes that are integrated in the analysis of the models in each chapter will encourage numerical experimentation, which in turn will lead to a deeper understanding of the economic intuition and limitation of these models. Our belief is that the synergy between the three central aspects of our approach – modelling, analysis and simulation – will enrich students' understanding of economic dynamics, their learning experiences and also help widen and deepen their skill set.

Since our main aim is to introduce economic dynamics to the undergraduate economics students, we have written it in a way that should make it accessible for the undergraduate programmes in economics and those specialising in economics in B.Comm. and B.A. programmes. The only prerequisite is to have some basic training in calculus, algebra and differential equations, which are usually part of modules on mathematics for economics and offered in the first year of the undergraduate programmes. Since the book arose from our joint module taught to the second year B.Sc. Financial Mathematics and Economics students between the academic years 2009–10 and 2019–20, it should be readily useful for such interdisciplinary undergraduate economics programmes. The book should also be of interest to other interdisciplinary science programmes where economics is introduced as part of the social sciences curricula. More broadly, the book should also be useful for general readers from the physical and mathematical sciences who are interested in the models of economic growth, inflation, or business cycles. Throughout the book, we have introduced concepts and historical notes, keeping such readers in mind, and we hope they find the book provides a useful introduction to the subject. Also the chapters in this book are purposefully kept short to cater to those readers from other scientific fields.

The book would not have been possible without the amazing and generous support of Routledge, particularly the senior editors during this project. The Covid-19 pandemic did throw the schedule and time lines in disarray, but the editors were extremely helpful, understanding and supportive through the project, and we owe them a great deal of gratitude.

The idea of the book and phases of its evolution happened in various periods and geographies. The book was one of our sabbatical projects and started in early 2019 in the temple city of Madurai in the southern Indian state of Tamil Nadu. We would like to extend our sincere thanks to the School of Mathematics in the Madurai Kamaraj University, especially to Prof. Lelis Thivagar, the Head of School, for giving us office space and other facilities, and other faculty members in the School, who welcomed us with warmth and friendship during our stay in Madurai, where the seeds for the book were planted. We would also like to thank Dr. Pushparaj in the School of Economics, who kindly helped us with the hosting arrangement in the School of Mathematics, and to Prof. Sankar Natesan, School of BioTechnology for his support during our stay. We would also like to thank Dr. Krishnakumar, Department of Physics at the Thiagarajar College in Madurai, for helping out and ensuring our stay in Madurai was very pleasant and also for his enthusiastic invitations for talks at his College.

We then moved to Austria, where we were invited to the *Institut für die Gesamtanalyse der Wirtschaft* (ICAE) at the Johannes Kepler University in Linz. We were very warmly welcomed by the young, friendly and dynamic colleagues at ICAE. We owe our debt of gratitude to Professor Dr. Jakob Kapeller, Director of the Institute, Dr. Stephen Pühringer, Dr. Katrin Hirte, Mr. Matthias Aistleitner, Dr. Claudius Gräbner and Dr. Bernhard Schütz. They made our stay very productive and enjoyable.

We would also like to acknowledge our home institution when the project started, with the School of Mathematics, Statistics and Applied Mathematics and the J.E. Cairnes School of Business and Economics, the National University of Ireland Galway, for facilitating our sabbatical leave and all the support for this project. We also like to acknowledge the research grant awarded to us towards the manuscript preparation by the Office of Vice President for Research at NUI Galway. Even though we did not use the money, we still would like to record our thanks to the Research office in NUI Galway for supporting this project. In the interim, Petri moved to Sweden and we would like to thank his current institution Chalmers Institute of Technology in Gothenborg, for facilitating his work for this book. We are also very greatful to MathWorks® who supported us through the *MathWorks® Book Program (software and promotion)* with MATLAB® access.

We would like to record our sincere thanks to all the B.Sc. Financial Mathematics and Economics students at NUI Galway between the academic years 2009–2010 and 2019–2020 who took the module *Modelling, Analysis and Simulation*. The idea of the book arose from the enthusiastic reception we received from the students across various years, who felt both challenged and engaged with the interdisciplinary module taught across the School of Mathematics, Statistics and Applied Mathematics and the School of Business and Economics. Running an interdisciplinary teaching module for a decade across two distinct administrative jurisdictions fundamentally requires continued interest and support of students, more than anything else. The book would not have been possible without their enthusiastic reception and valuable feedback on the module.

In the same vein, we would also take this opportunity to record our sincere thanks to all our teachers, who opened the world for us and without whom this book would not have been possible.

Last but not the least, we thank our families, who were behind us and motivating us every step of the way – Thank you!

1 INTRODUCTION TO ECONOMIC DYNAMICS

The modern economy is a complex system comprising heterogeneous agents such as individuals, firms, and governments interacting with each other directly and indirectly in the market place. The main aim of Economics, as a subject of study in the Social Sciences, is to understand the workings of the economies and to explain the underlying mechanisms that drive such complex systems. Classical economists, such as Adam Smith, who is considered as the father of economics, emphasised the role of an individual agents' self-interest in seeking their own gain in the market as the mechanism driving the self-regulating nature of the modern economies, as if regulated by an *invisible hand.* In contrast, Karl Marx argued that modern economies are characterised by the inherent tendency for unequal exchange between the owners of capital and the workers in the marketplace and consequently the economic relations between the agents that underpin the structure of production and determine distribution and economic growth in such complex systems. These contrasting views about the conception of modern economies underpin and continue to inform the development and articulation of contenting economic policy paradigms.

Economic theory strives to understand the mechanisms that underpin the workings of the modern economies through *models.* An economic model is an abstract description of a particular economic reality, since the complexity of the reality under investigation cannot be comprehended in its entirety. The subjective nature of scientific exploration in economics arises from our conception of the economic system, which is driven by a host of overlapping socio-cultural-political factors, and as a result there exist multiple and differing perspectives or conceptual frameworks. Economists often agree upon a set of empirical realities, known as *stylised facts*, and try to explain those realities or outcomes from their own point of view. For example, there is a rich diversity of economic models that explores the questions of unemployment, inflation and economic growth. The fundamental purpose of *theoretical models* is to elicit features of the underlying, or generative mechanisms that

DOI: 10.4324/9780429324406-1

generate those particular stylised facts. In contrast, *empirical models* aim to verify those theoretical propositions about the generative mechanisms of the phenomenon under investigation.

Therefore, the term *modelling* in economics refers to deriving *meaningful* hypotheses about an economic phenomenon or reality of interest. A meaningful hypothesis proposes plausible explanation of the specific phenomenon under investigation and lends itself for verifiability where it can be validated or refuted. However, what is *meaningful* depends on the particular conceptual framework, or lens, that is being used to view the system. In this sense, modelling in economics is influenced by the particular conceptual framework that is being used to view the empirical phenomenon. This is one of the reasons why there are several "schools of thought" in economics, each providing distinct explanations for economic phenomena such as unemployment, inequality, growth etc. It is this particular characteristic that distinguishes economics from the pure science disciplines (mathematics, physics, chemistry, etc.).

Perspectives also influence the mode of analysis such as if an economic phenomenon is to be investigated in a *static* or a *dynamic* mode. The static mode is one where the notion of time is suspended, as if one is analysing a snapshot of the system, and the analysis focuses on the relationships between variables in that timeless context. If the analysis, on the other hand, is concerned with the evolution of the relationships over time then time needs to be explicitly introduced. The latter mode of analysis is done through specifying dynamic models of the phenomenon under study. Consider, for instance, the textbook model of demand and supply (see Chapter 2), where the equilibrium price (and quantity) is determined by what is demanded and what supplied, or where the demand curve intersects the supply curve. In determining the equilibrium price (and quantity) in a static mode, the only consideration is the equality between demand and supply, and there is no reference to the passage of time. In other words, the static analysis looks at the relationships, say between demand and supply, at a specific point in time. When we compare two static equilibrium positions the analysis is known as *comparative statics*.

In contrast to static analysis, dynamic analysis is concerned with the description of movement from an arbitrary out-of-equilibrium position to the equilibrium, and between one equilibrium and another. In other words, the dynamic analysis considers the notion of time explicitly. Answers to such questions is not just theoretical curiosity but imperative for practical policy making. In fact, in many instances practitioners and policy makers require an understanding of *how* the market is going to reach its equilibrium, and also *how long* it will take to reach the equilibrium. In the context of the

Covid-19 pandemic, policy makers were concerned with whether the nature of the economic recovery would be a "K- or W-shaped" recovery and the policy measures to alleviate such an unequal recovery process. However, for making such predictions one needs to provide a meaningful description of the evolution of the system over time and dynamic analysis provides such a description. Even for the comparative statics analysis to yield meaningful results it is necessary to understand the dynamics of the system under study. Paul Samuelson, a Nobel Memorial Prize winner in Economics, reminded us about the importance of dynamic analysis in his seminal book *Foundations of Economic Analysis* published in 1947:

> In order for the comparative-statics analysis to yield fruitful results, we must first develop a theory of dynamics.
>
> (Samuelson [3], pp. 262–63)

In the comparative statics analysis there is an assumption that the *equilibrium* of a model is where the system settles down and to which it will return after a small perturbation. However, this assumption need not always hold. Consider for instance the equilibrium price in the demand and supply model, which is obtained at the intersection between demand and supply functions. The assumption is that the equilibrium price (and quantity) is where the system will settle down. However, this is true only under certain conditions. In the demand and supply model, as you will see in Chapter 2, the system settling down at the equilibrium point depends on the relative slopes of the demand and supply functions. If the relative slopes are not favourable, the system may never gravitate towards the equilibrium price. Therefore, we need to qualify the *nature* of the equilibrium. It is a common practice in Economics to use the notion of equilibrium without any qualification, whether they are stable or not. Therefore, in the dynamic analysis we first need to determine the *stability* of the equilibrium and then quantify whether it is a *stable* or an *unstable* equilibrium. This is where the study of economic dynamics begins.

Understanding the characteristics of an equilibrium, i.e. its stability, is important in the study of the nature of movement towards or away from the equilibrium position. This crucial insight is often missed in the static and comparative static analysis. The analysis of the stability of an equilibrium will, however, reveal the characteristic features of the equilibrium that one is dealing with. How does the system converge to a particular value, i.e. an *equilibrium point?* Does it converge *monotonically* or does it exhibit *oscillatory convergence*? Does it *oscillate* around the equilibrium value? Characterizing the nature of the oscillations are essential when studying aperiodic fluctuations, e.g. volatility in the stock market, and periodic fluctuations, e.g. business cycles that are often exhibited by the economic data. For instance, in the demand-supply model, we are interested in whether the market tends to a particular price in a monotonic way or exhibits oscillations before converging. These questions are important for our understanding of how a system, e.g. the economy, would behave if perturbed by a *shock,* like the Covid-19 pandemic, and its movement back towards the equilibrium position.

There is also the question of movement between different equilibrium positions. For instance, in the demand and supply model, if there is a shift in the demand curve, due to a change in the price of the substitute good, then the system moves to the new equilibrium position. In comparative statics analysis the two equilibria are compared from a welfare perspective. However, the characteristic of the new equilibrium and the movement between the equilibria is not taken into account in the welfare analysis. Suppose the new equilibrium is stable but oscillatory, then the nature of convergence would be qualitatively different from the dynamics around the previous equilibrium. On the other hand, if it is unstable then the market never reaches the new equilibrium position, i.e. the market never *clears.*

These two questions, namely, the characteristic feature of an equilibrium and the nature of movement between equilibrium positions, are quite fundamental for our understanding of the economic system. Exploration of these questions provides us with an understanding of the fundamental mechanisms that underpin the economic phenomena in modern economies. It is akin to asking what is the dynamics (mechanisms) that generates earthquakes, for instance. In economics, we are interested in the dynamics that generates booms and busts, economic growth and the disparities in economic growth between different economies, inflation and unemployment etc. Since we are trying to understand these phenomena using economic data, which is generated by the dynamical system called *the economy*, we need to use the dynamical-systems approach that can shed light on the evolution and other dynamical properties of the economic system.

This book aims to provide a gentle introduction to these concepts using some well known models in Economics. Most of these models are introduced in the first and second year undergraduate modules and some in the senior undergraduate modules. Mostly, these models are taught in the comparative statics mode and students do not always get to see the dynamics that underpin such analysis. Our aim in this book is to explore the dynamics that underpin these models and help students better appreciate the economic intuition of these models. To that end, we use the economic logic of the model to guide us and lead us to the dynamic analysis, rather than the other way around. We believe this approach will keep the economic intuition or logic in the forefront of the analysis with the hope that students do not feel overwhelmed by mathematical modelling. There are some excellent books that provide rigorous introduction to economic dynamics at this level, for instance, by Giancarlo Gandolfo [2] and Ronald Shone [4]. In relation to these text books, our objective here is more modest. We aim to initiate students to think in terms of dynamics, and prepare them to explore using simulation methods, using some of the popular economic models taught in the undergraduate modules such as microeconomics and macroeconomics. This, we hope, will create sufficient impetus for learning economic dynamics and help students to pursue more advanced books such as by Gandolfo, Shone and others in this area.

In this book, in addition to initiating students' interest in dynamics, we also aim to introduce them to simulation methods using the MATLAB® programming language. In our view, and from our own experience of teaching economics in Ireland and in various other countries, there is a lack of computational training at the undergraduate level in economics programs. More often than not, students of economics miss out on this crucial skill that would help them to prepare for the so called fourth industrial revolution. Here we introduce students to simulation and visualisation, which help students to analyse, visualise and experiment with models. We believe that this is another distinguishing aspect of the book and we hope it will induce an interest in working with programming languages, in this instance MATLAB® and will give students confidence to pursue others such as R and PYTHON, all of which have become industry standard and widely used in the banking, finance/fintech sectors, and in economic and public policy institutes. While the programming languages like MATLAB® have a relatively steeper learning curve than the worksheet-based programs like MS Excel, we believe that the long-term benefits more than outweigh the initial time spent on learning the language. There are many primers and introductory books on MATLAB® computing. For our purposes, we provide a very brief introduction to MATLAB® in the following section.

1.1 A SHORT PRIMER ON MATLAB®

The software known as MATLAB® (for "MATrix LABoratory") is an extraordinarily useful computational tool for analysis and design, which we will use for numerical analysis and visualisation throughout this book. Here we will provide a very short introduction to the language of MATLAB®, which is not meant to be exhaustive or complete but to give a taste of the language.

Start the MATLAB® program by clicking on the MATLAB® symbol in your computer where-after the MATLAB® command window will appear. You will see that the window has several panes and we will initially be entering commands into the "Command Window". Defined variables appear in the "Workspace window" (upper left side) and previous commands appear in the "Command History" window (lower left).

It is possible to use the command window as a normal calculator. We can for instance perform basic calculations.

Basic calculations

```
>> 2+4                         >> 2^4
ans =                          ans =
     6                              16
```

Seeking information about functions is easy in MATLAB®.

Elementary functions

Type

```
>> help
```

in the command window and you will get a list of useful functions, such as

`sin, cos, tan, exp, abs, log, sqrt, round`

It is also possible to write

```
>> help ops
>> doc ops
```

which gives you a list of useful operators, such as

`+, -, *, ^, /, \, .*, ./`

The command `doc` provides comprehensive information about the query.

There are a fair number of special operators, variables and constants in MATLAB®

Special variables and constants

Some operators

`ans`	Most recent answer.	`Inf`	Infinity
`eps`	Floating-point relative accuracy.	`NaN`	Not a number
`i,j`	Imaginary unit.	`pi`	3.141592653589793

Time

`clock`	Wall clock.	`etime`	Elapsed time function.
`cputime`	Ellapse CPU time.	`tic, toc`	Stopwatch timer function
`date`	Calendar.		

Every entity in MATLAB® is represented as a matrix. Therefore, a good understanding of elementary operation of vectors and matrices would be very helpful.

Vectors and matrices

Create a row vector v with elements 1,3,4,6

```
>> v = [1,3,4,6]
v =
      1     3     4     6
```

and create a column vector of v with elements 1,3,4,6 in two ways

```
>> v1 = [1;3;4;6]
v1 =
       1
       3
       4
       6
```

```
>> w = v'
w =
       1
       3
       4
       6
```

Notice that the *transpose* operator is given by '. Next, create a row vector v with elements 1,2,...,7 by

```
>> v = 1:7
v =
      1     2     3     4     5     6     7
```

and create a row vector v with elements 1.1,1.3,..., 1.9 using :

```
>> v = 1.1:0.2:1.9
v =
    1.1000    1.3000    1.5000    1.7000    1.9000
```

Creating matrices and manipulating matrices is very easy to do, but there are many different ways to do it. Here a few of them are shown.

Vectors and Matrices

To create a 2 × 3 matrix M with elements 1,3,4,6,7,9 write either

```
>> M = [1,3,4;6,7,9]
M =
       1     3     4
       6     7     9
```

or

```
>> M =   [1 3 4
          6 7 9]
M =
       1     3     4
       6     7     9
```

or

```
>> M(1,1) = 1; M(1,2) = 3; M(1,3) = 4;
   M(2,1) = 6; M(2,2) = 7; M(2,3) = 9;
>> M
M =
       1     3     4
       6     7     9
```

or

```
>> M(1,:) = [1,3,4]; M(2,:) = [6,7,9];
>> M
M =
       1     3     4
       6     7     9
```

or

```
>> M(:,1) = [1;6]; M(:,2) = [3;7]; M(:,3) = [4;9];
>> M
M =
       1     3     4
       6     7     9
```

To create a 3 × 3 identity matrix Id write

```
>> Id = eye(3)
Id =
       1     0     0
       0     1     0
       0     0     1
```

MATLAB® is as mathematical and scientific computing tool, not only a calculator, and offers a lot of flexibility. One of the main features of MATLAB® is the use of **M-files**, which are simply files with a set of code to perform a computation that can be saved, instead of typing directly in the command window, and recalled in any other program. M-files can be either **script files** that have no input or output arguments; or **functions** that have input arguments. They are useful for automating a series of MATLAB® commands, such as computations that you have to perform repeatedly from the command line. The M-files are executable files and they are saved in MATLAB® with the extension `.m`.

M-files

To open a new file choose File → New → M-File from the Menu bar and start typing your MATLAB® code. As always, **Don't forget to save!!**, but files will be automatically saved when executing an .m-file.

```
x = linspace(0,2,100);                % variable vector x
                                      % with 100 elements.
y = sin(2*pi*x);                      % output

plot(x,y,'k','LineWidth',2)           % Plot of x versus y
axis([0 2 -2 2])                      % Set figure axes

xlabel('x'), ylabel('y')              % Name of axes
title('My first plot')                % The figure title
```

Save this as the M-file `myfirstplot.m`.

Then, in the command window type following command

Execute the m file

```
>>myfirstplot
```

and you will get the output in Figure 1.1.

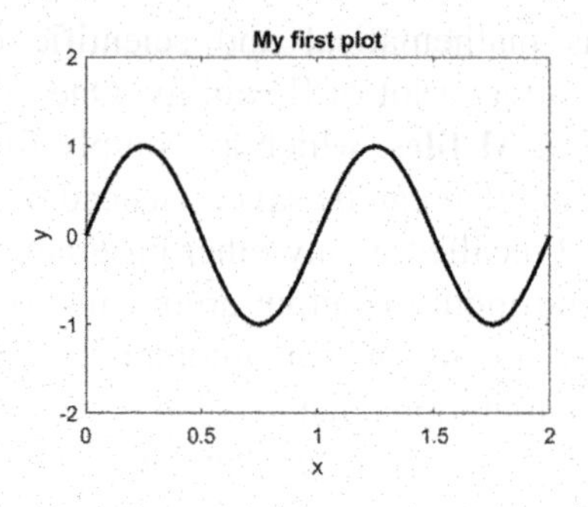

Figure 1.1: The graph generated by the file `myfirstplot.m`.

Plotting graphs is very important for anybody working in any applied sciences and MATLAB® is very good for this purpose.

The built-in MATLAB® function `PLOT(X,Y,S)` plots vector X versus vector Y, where S is a character string describing the way in which the data is shown. If X or Y is a matrix, then the vector is plotted versus the rows or columns of the matrix, whichever line up. If X is a scalar and Y is a vector, disconnected line objects are created and plotted as discrete points vertically at X.

To find lots of information on how to plot your data, type `help plot` or `doc plot`.

Visualisation

Consider the following example where we will graph three different functions and show it in the same plot

```
x = 0:4*pi/100:4*pi;        % variable vector x with 100
                            % elements.
y1 = sin(x);                % output of the first function
y2 = 2*sin(x);              % output of the second function
y3 = sin(2*x);              % output of the third function

figure(1)                   % open a new figure window
                            % called 1
hold on                     % do not erase old plots when
                            % plotting
% Plot of x versus y1; solid black (k); width 2cm
plot(x,y1,'k','LineWidth',2)
```

```
% Plot of x versus y2; dashed red (r--);
plot(x,y2,'r--','LineWidth',2)
% Plot of x versus y3; dotted blue (r:);
plot(x,y3,'b:','LineWidth',2)
% label the axes
xlabel('x'),ylabel ('y_1, y_2, y_3')
axis([0 4*pi -2.5 2.5]) % set figure axes
box on                  % outline box on
hold off                % erase old plots when plotting
```

Save this as the M-file `mysecondplot.m`. The output of the above code is displayed in Figure 1.2.

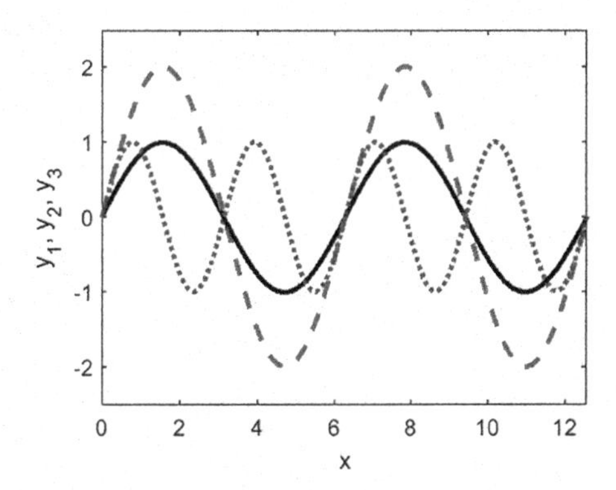

Figure 1.2: The graph generated by the file `mysecondplot.m`.

While the above discussion provides a brief introduction to MATLAB® programming, one of the issues with such an approach is similar to the problems associated with the rote learning strategies in mastering a language. The rote learning strategies use memorisation, repetition and practising of vocabulary and grammar, which may be helpful in the short term, but a more effective way to mastering languages is through contextual learning. Linguistic scholars, like Henry Douglas Brown [1], have long argued that by learning how to use words in different contexts the learner is more sure of its meaning and such an approach will also help the long-term memory of those words.

In this book, we follow the contextual approach in introducing MATLAB® to students who are new to programming. Since the economic problem provides the background context for MATLAB® programming, we believe that students will see the purpose, usefulness of simulation and visualisation, which will enable them to experiment with models and provide an avenue for experiential learning.

In the following example, we illustrate our approach using the basic *Demand and Supply model.* This model will be examined at length in Chapter 2. The purpose here is to use the model as an illustration for simulation and visualisation in MATLAB®.

Simulation & Visualisation: An economic example

The most basic model of demand and supply equation is given by

$$D_t = d_0 + d_1 p_t, \quad b < 0, \tag{1.1}$$
$$S_t = s_0 + s_1 p_{t-1}, \quad b_1 > 0. \tag{1.2}$$

The equilibrium condition $D_t = S_t$ yields the first-order difference equation

$$d_0 p_t - d_1 p_{t-1} = s_0 - s_1 1 \quad \Leftrightarrow \quad p_t = \frac{s_1}{d_1} p_{t-1} + \frac{s_0 - d_0}{d_1}, \tag{1.3}$$

which describes the evolution of price p from one time period to the next. Setting the parameter values of the demand and supply functions and the initial value p_0 of price to

$$d_0 = 4, \quad d_1 = -1.7, \quad s_0 = 0.5, \quad p_0 = 1,$$

yield the simulation plots in Figure 1.3 (a), for $s_1 = 1.1$, and Figure 1.3 (b), for $s_1 = 1.9$. See Chapter 2 for a full exploration of this example.

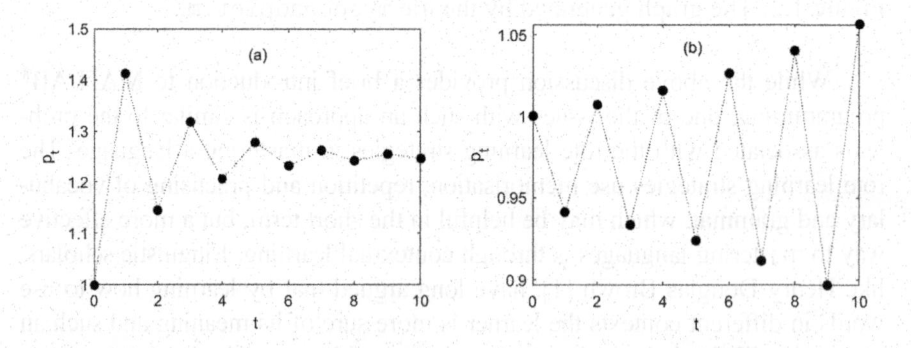

Figure 1.3: Evolution of price p_t over time t. In (a) $s_1 = 1.1$ and the price converges in an oscillatory manner and in (b) $s_1 = 1.9$ the price diverges in an oscillatory manner.

Using the MATLAB® code corresponding to the above example (see Chapter 2 and Appendix B), students can experiment with the parameter values for demand and supply functions, and also varying initial conditions, and visualise the model in various scenarios. Such an experimentation helps students to see when the models turn from having a stable equilibrium to an unstable equilibrium, and understand the economic intuition behind those scenarios. Thus, the MATLAB® simulation enables experimentation and allows students to understand the underpinning dynamics of the model by simply varying the parameters of the model before they even learn to analytically solve the model. As students begin to see the visual output of their simulation in terms of plots and begin to analyse them, learning through experimentation will begin to create a self-reinforcing effect in terms of their confidence and interest.

We believe that the context-based learning approach adopted in this book, where economic problems lead and motivate the dynamic analysis and simulation using MATLAB®, helps not only in introducing a new programming language but also helps initiating the undergraduate students to take up the study of dynamical systems. The interwoven nature of discussion between modelling, analysis and simulation is a unique feature of this book and we hope it will help bring the joy of experiential learning to economics classrooms.

BIBLIOGRAPHY

[1] Brown, D.H. (2014), *Principles of Language Learning and Teaching* (6th Edition), Pearson Education ESL.

[2] Gandolfo, G. (1997), *Economic Dynamics,* Springer Science & Business Media.

[3] Samuelson, P.A. (1947), *Foundations of Economic Analysis,* Harvard University Press.

[4] Shone, R. (2001), *An Introduction to Economic Dynamics,* Cambridge University Press.

2 THE COBWEB MODEL

2.1 ECONOMIC PROBLEM

Markets are the fundamental organizing institutions of a modern economy. Every market, be it for potatoes, labour or financial assets, enables the coordination of decentralised decision making by bringing together disparate buyers and sellers of goods and services. The study of individual markets and the decision – making process of buyers and sellers fall in the realm of *microeconomics*. The traditional neoclassical approach describes the decision-making process of buyers and sellers such as households, firms or governments using the analytical framework of utility or profit maximisation.[1] The description rests on various assumptions such as perfect competition, i.e. atomistic individuals, perfect knowledge or rational behavior, and no transaction costs or friction-less markets etc.

At one level the success of neoclassical theory can be attributed to its uncompromising focus on scarcity, and hence competition as the key to understanding economic behaviour, as evinced in microeconomics. However, the analytical formalism and elegance of microeconomic theory comes at the cost of using some of the most unrealistic assumptions such as rationality, which underlies the frictionless conception of the economy. The rational-choice framework assumes that individuals *know* the underlying model of the economy and are able to discern what is best for them in their self interest and act accordingly. Such a decision-making process fails under the conditions of uncertainty, where we cannot possibly know what we need to know in order to maximise, or even safe guard our self interest. Herbert Simon [2] articulated this more precisely:

> If ... we accept the proposition that both the knowledge and the computational power of the decision maker are severely limited, then

DOI: 10.4324/9780429324406-2

> we must distinguish between the real world and the actor's perception of it and reasoning about it. That is to say we must construct a theory (and test it empirically) of the process of decision. Our theory must include not only the reasoning processes but also the processes generated the actor's subjective representation of the decision problem, his or her frame.

The analytical elegance, the mathematical formalism and the precision of microeconomic theory also comes with another fundamental limitation, namely, that it is set in a timeless, or static, conceptual structure. As Douglas C. North [3] put it more eloquently:

> It is not clear where economics is going. But the direction is suggested by two glaring shortcomings of neoclassical theory: it is a frictionless theory in a world in which frictions are where the action is, and it is static in a world in which dynamic change is going on at an unprecedented rate.

The power of microeconomic theory comes from its simplicity in organizing economic behaviour of people in terms of two fundamental precepts, one at the individual level and the other at the market level. On the individual level economic decisions of people can be understood by the *optimisation process,* e.g. people try to make the best possible choices they can afford. On the market level the functioning of markets can be understood through the *equilibrium process* where, for instance, prices adjust dynamically and coordinate buyers and sellers to trade goods and services that benefit both sides.

Of the two, the assumption that economic agents use the process of optimisation to arrive at the best possible outcomes for themselves is relatively less problematic, although it assumes certain pre-suppositions such as agents have the freedom to choose or that they have complete information etc. However, the notion of an equilibrium, as an analytical mechanism to understand how markets function, needs careful consideration since the meaning of an "equilibrium" in an economic context is not an easy term to define precisely. Since we are concerned with economic dynamics, let us first discuss the notion of *equilibrium* and how it is applied in economics.

The definition of the term equilibrium differs depending on the context of the problem. In economics, we define the term as a state of *rest,* which could be a fixed value of a variable that has no tendency to change – like a rocking horse that "settles down" to the position of rest after being rocked.

In physical systems, an equilibrium is often defined as a state of *balance* between opposing forces. For example, if the state of a mechanical system is *at rest* or in a *non-accelerated motion* then the total sum of all forces acting on the system is zero. We can also have a situation where a system exhibits a repeated pattern without ever coming to a single point of rest, but we still refer to such a behaviour as a *steady-state* motion. For example, think of a central air-conditioner that is repeatedly switched on or off such that the room temperature oscillates between two different values and stays within a desired band. The *steadily* oscillating room temperature can be seen as "steady-state path", as long as there are no other external disturbances like power failure etc.

In this book, we will encounter all these types of equilibria or steady-state behaviours, and in this chapter we will study the first type of equilibrium, one where the system, or the market in question, "settles down" to the position of rest. However, even if a system has an "equilibrium" or "steady state" it does not automatically mean that the system settles down to a particular value, or a particular pattern. It is one thing for a system to possess an equilibrium, but it is another thing whether the system, starting from any arbitrary point away from equilibrium, is able to get to it, i.e., whether it is *stable* or not. The stability of an equilibrium point is crucial because it determines whether the system, starting from a nearby initial condition, converges towards the equilibrium value or diverges away from it. For instance, a particular market may have an equilibrium price (and quantity) at which it clears, where demand equals supply. But, mere existence of such an equilibrium price by itself does not guarantee that starting from an arbitrary initial price, the market will converge to that price. Only if the equilibrium is stable, the market converges to the market clearing or the equilibrium price. However, if the equilibrium is unstable, the deviation between the demand and supply price grows over time leaving the market negotiations to diverge and the market does not clear. Therefore, in addition to determining equilibrium values of particular markets, it is crucial to determine whether they are stable or not. In other words, the term 'equilibrium' needs to be qualified with respect to whether it is a stable or unstable equilibrium.[2]

Although much less appreciated in the undergraduate teaching, the notion of stability of equilibria is fundamental to economics, be it in microeconomics or macroeconomics. The fundamental question here is whether the system will return to its equilibrium value after being perturbed away from it. If the answer is affirmative then there are two further questions that need to be answered. First, what is the nature of the convergence, i.e. will it converge back to the equilibrium in a monotonic or oscillatory manner? Second, how long will it take for the system to return to its equilibrium value? The importance of such questions can for instance be seen in the context of the Covid-19 crisis, where the economic policy discussions were focussed on whether the economy will experience a "V-", "W-", or "K"-shaped recovery and what would be the time-scale (or speed) of the recovery process.

In this chapter we will study this question in a microeconomic setting using a basic demand and supply model, referred to as the cobweb model. Every student of economics is introduced to this basic demand-and-supply model, or the *Marshallian scissors*, and remains as a basic pedagogic tool for the analysis of markets. Here, we consider a single market, as shown in Figure 2.1, where the relationship between the price (p) and quantity (q) of a product are given by demand and supply curves labelled as D and S, respectively. We assume, for simplicity, that demand and supply depend linearly on the price. Suppose the initially quoted price for the product is p_0, at which there is excess demand, i.e. the quantity demanded by the consumers is greater than the quantity offered by the suppliers. The usual analysis of the problem would argue that excess demand would drive the prices upwards because the suppliers notice that there is demand for their product and expect

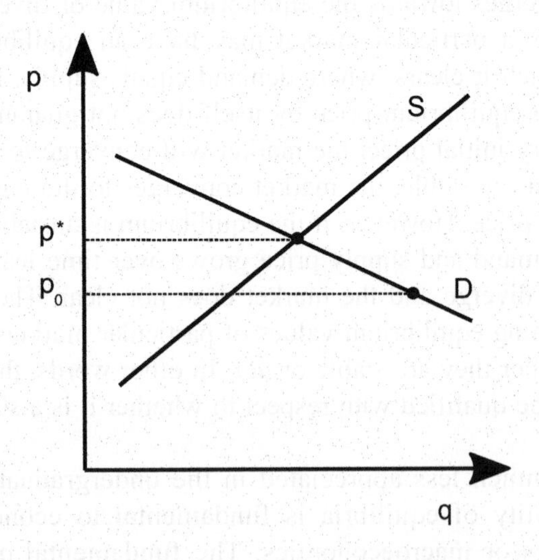

Figure 2.1: A schematic representation of a demand-and-supply model for a product, where the quantity q is plotted versus the price p with D representing demand and S representing supply. An initial price p_0 as well as the equilibrium price p^* are indicated.

they could sell the corresponding quantity at that higher price, as shown by the supply curve S in Figure 2.1. However, at that higher offer price, buyers would demand a lower quantity, as dictated by the demand curve D, than that offered by the seller. This process of *tâtonnement*, or grouping orders, will continue until the market reaches the equilibrium price p^* and the corresponding equilibrium quantity q^*, where all the excess demand is eliminated.

In the above analysis, two things are not discussed. Firstly, what is the guarantee that the process will converge to the equilibrium price? One should be aware of the possibility that starting from an initial price p_0, the price may diverge away from the equilibrium price, i.e. the negotiation between the buyer and seller could break down and that no trade takes place. Secondly, assuming that the market converges to the equilibrium price and quantity, how does it converge, i.e. does it converge *monotonically* or *oscillatory*?

What we are attempting to describe here is the *disequilibrium dynamics* of the price mechanism in this market, i.e. what makes the price move from p_0 to p^* and the characteristic features of such a movement. Such a description is fundamental to both theoretical and policy analysis, because we need to know what renders markets unstable to be able to prescribe effective stabilisation policies. In the following section, we consider the basic cobweb model of demand and supply in a single market, and explore these questions. We will use simulation and visualisation techniques to highlight the underpinning dynamics as well as analyse how the model behaves under parametric variations.

2.2 MODELLING

Consider a linear demand and supply model in *discrete* time[3] given by

$$D_t = d_0 + d_1 p_t, \qquad d_0 \geq 0, d_1 < 0, \tag{2.1}$$
$$S_t = s_0 + s_1 p_{t-1}, \qquad s_0 \geq 0, s_1 > 0, \tag{2.2}$$

for $t = 1,2,3,\ldots$ and some initial price p_0, where D_t represents the demand at time t as a function of the price at time t (p_t) and S_t represents the supply at time t as a function of price at time $t-1$ (p_{t-1}). The intercept and slope parameters d_0, d_1, s_0 and s_1 are specific to the goods being bought and sold. Here we are dealing with goods whose demand is negatively related to price p_t in the current time t, i.e. $d_1 < 0$. On the other hand, since supply of the produce is positively related to the price p_{t-1} in the previous time $t-1$, the slope of the supply function is positive, and so $s_1 > 0$. In Figures 2.2 (a) and (b) schematic representations of the demand and supply curves D and S are shown and in particular we note that D depends on p_t and S on p_{t-1}.

Historical note

The original cobweb model was developed in the context of agricultural markets where the supply of produce is influenced by the time lag between planting and harvesting, and where the producer's price *expectations* are assumed to be based on a previous period's price realisation, see Nicholas Kaldor [4].

The economic intuition behind the supply curve S_t is that the suppliers (e.g. farmers) would supply more of their produce in the present time t, if they had received a favourable price at the previous time period $t - 1$. In Figure 2.2 (c) we see two steps of how the negotiation process works, which is driven by the fact that the demand and supply curves depend on the price at different times, and will be discussed further in Section 2.3.

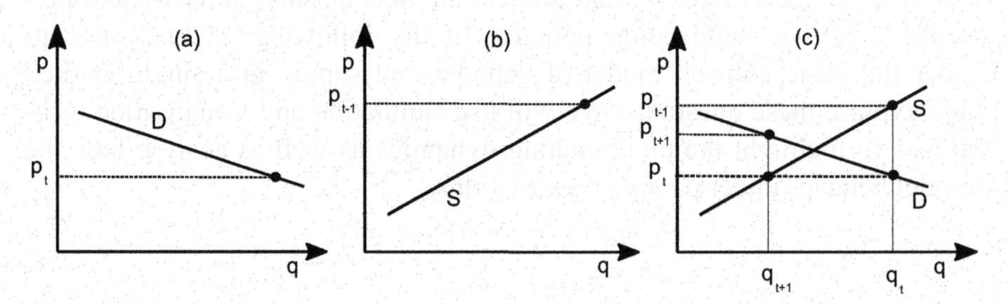

Figure 2.2: Schematic representation of the demand and supply model, where in (a) the demand curve (2.1) and (b) the supply curve (2.2) are shown. In (c) a representation of two steps of the negotiation process between a buyer and seller is shown.

Finally, the situation where both the consumer and supplier are satisfied in an exchange referred to as the *market clearing condition* and which happens when

$$D_t = S_t. \tag{2.3}$$

Specifically note that the *equilibrium price* is given by the market clearing condition (2.3) and is represented by p^* in Figure 2.1. We shall now proceed to study the determination of the equilibrium price (and the corresponding quantity q^*) and analyse its stability properties in the next section.

2.3 ANALYSIS, SIMULATION AND VISUALISATION

Let us analyse the demand-and-supply model set up in Section 2.2 by substituting equations (2.1) and (2.2) into the market clearing condition (2.3) that yields

$$d_0 + d_1 p_t = s_0 + s_1 p_{t-1}, \quad t = 1,2,3,\ldots \tag{2.4}$$

or

$$p_t = \frac{s_1}{d_1} p_{t-1} + \frac{s_0 - d_0}{d_1}, \quad t = 1,2,3\ldots \tag{2.5}$$

which is a *first-order difference equation* that gives us a relationship between the price p at time t and at $t-1$.

In order to *iterate* the equation we need an *initial condition,* which is an "initial" price at a specific time period. For instance, let p_0 be the initial price at $t = 0$ (see Figure 2.1). Once we have specified the initial price, we can use it in (2.5) and see how the price evolves, i.e. $p_0, p_1, p_2, \ldots$, as a result of the negotiation between the buyer and the seller.

In the traditional treatment of the demand-and-supply models one typically solves the first-order difference equation (2.5) and use the analytical solution to describe the dynamics. However, we shall take a more qualitative and experimental route by using simulation of (2.5) in MATLAB® to get an intuitive feel for the underlying dynamics of the model. The following MATLAB® code sets up the parameter values and the initial condition.[4]

Parameter values and initial conditions

In order to start a simulation, we need to specify numerical values for the parameters d_0, d_1, s_0 and s_1 as well as the initial price p_0. In what follows we will use the following numerical values guided only by the sign restrictions

$$d_0 = 4, \quad d_1 = -1.7, \quad s_0 = 0.5, \quad s_1 = 1.1 \quad \text{and } p_0 = 1.$$

Note that these values are only used as an example. You are encouraged to experiment with different parameter values (even "non-allowed" or unrealistic values) in order to get a better feeling for the dynamics of the system.

Initiation and parameter values

```
%Preamble
clear all % Clears all variables
close all % Closes all figure windows

%Simulation parameters - specifying parameter values
d0 = 4;
d1 = -1.7;
s0 = 0.5;
s1 = 1.1;

%Initial Condition
p0 = 1;
```

Next we need to specify the length of time we would like to iterate the equation (2.5). For our purposes, we iterate the equation for 10 time steps, and the code to stipulate the time steps is given as follows.

Simulation parameter

```
%Simulation parameter - specifying the number of
%iterations
n=10;
```

We can now iterate the difference equation (2.5) using the following code.

Simulation

```
%Simulation
p=[];
p(1) = p0;
for t = 1:n
   p(t+1) = (s1/d1)*p(t) + (s0-d0)/d1;
end
```

To briefly explain the code, we first create an empty vector p where the first element p(1) is being equal to the initial condition p0. Note, the first index of a vector in MATLAB® is 1 so, for instance, `p(0)` is not allowed.

After initialising p, we iterate the difference equation (2.5) for $n = 10$ time steps using a for-loop, where t is the index representing time. The end result will be stored in p, which is a vector of length 11 with entries $p_0, \ldots, p_{10}$.

When the simulation (the for-loop) is completed, we can visualise the output to see the evolution of the price p_t with respect to time t. We use the following code to visualise the price dynamics:

Visualisation

```
%Visualise time versus price
figure(1)
plot(0:n,p,'ko','MarkerFaceColor','k')
xlabel('t'),ylabel('p_t')
```

In the above set of visualisation commands, we first open an empty figure pane, called "Figure 1". We then plot the time vector 0:10 = [0, 1,..., 10] versus the price vector p = [p(1),..., p(11)] and label the horizontal axis t and the vertical axis p_t. Figure 2.3 shows the output from this set of commands.

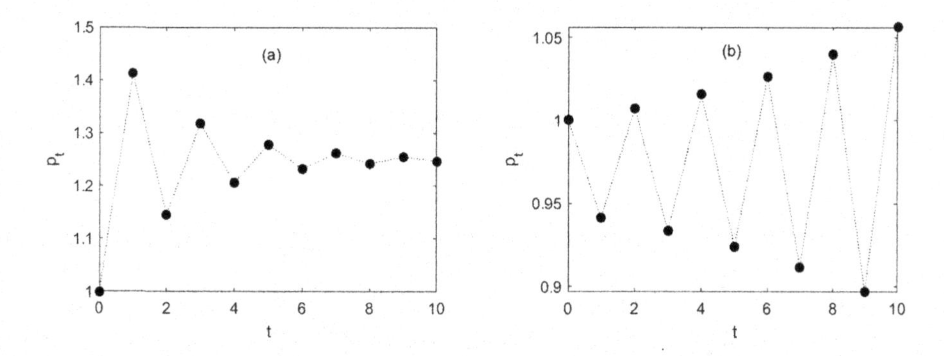

Figure 2.3: Evolution of price p_t over time t with initial condition $p_0 = 1$. In (a) $s_1 = 1.1$ and the price converges in an oscillatory manner and in (b) $s_1 = 1.9$ and the price diverges in an oscillatory manner.

In Figure 2.3, we can see two scenarios emerging in this model. In Figure 2.3(a), we set $s_1 = 1.1$ and see that starting from the initial value of $p_0 = 1$, the model yields oscillations in price, but these oscillations gradually dampen and converge to a fixed value around $p = 1.25$. We can also verify this by simulating the model for longer duration by increasing the value of n. Next, in Figure 2.3(b) we set $s_1 = 1.9$ and see that starting from the same initial value, the model yields a diverging price sequence as the amplitude of the oscillation increases over time.

Question

Why does the model produce two different price dynamical scenarios? In one case (Figure 2.3(a)) the price converges to a fixed value in an oscillating manner and in the other case (Figure 2.3(b)) the price diverges without bound in an oscillating manner.

To answer this question, let us first *analytically* solve the difference equation (2.5). We start by solving for the fixed or *equilibrium price* (p^*) that should exist, as suggested by the simulations in Figure 2.3.[5] Since the price sequence converges to a particular fixed value (p^*), we solve for the equilibrium price by substituting $p_t = p_{t-1} = p^*$ in (2.4), which yields

$$d_0 + d_1 p^* = s_0 + s_1 p^*, \tag{2.6}$$

and so the *equilibrium price* in this market is

$$p^* = \frac{s_0 - d_0}{d_1 - s_1}. \tag{2.7}$$

The equilibrium price is also called the market-clearing price as it is the price at which the demand equals supply and the market clears, as seen in Figure 2.4.[6] We also note that $d_1 - s_1 \neq 0$ (see (2.1) and (2.1)), which means that the equilibrium price (2.7) is well-defined and economically meaningful. From (2.1), (2.2) and (2.7) we can calculate the equilibrium quantity q^*, corresponding to the equilibrium price p^*, that is

$$q^* = \frac{d_1 s_0 - d_0 s_1}{d_1 - s_1}. \tag{2.8}$$

Next we want to study the stability properties of the market equilibrium, or in other words, we want to determine under what conditions the

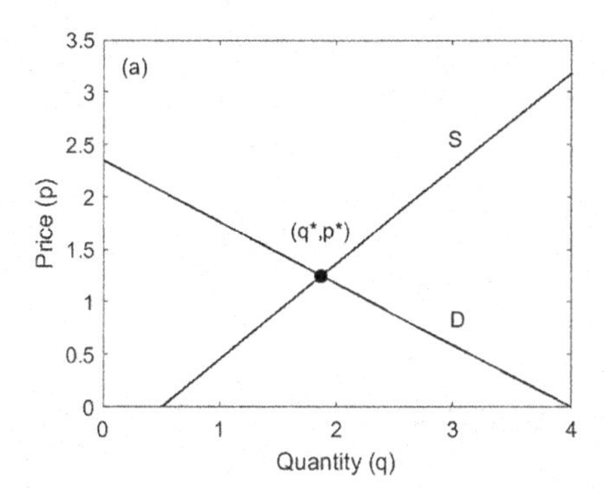

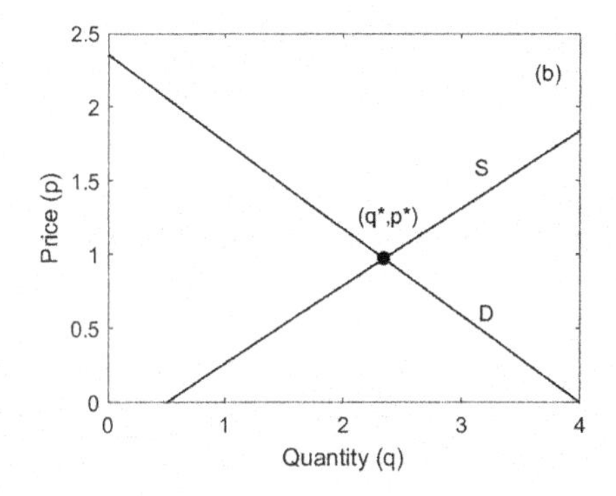

Figure 2.4: Quantity q versus price p. In (a) $s_1 = 1.1$ and in (b) $s_1 = 1.9$. The demand and supply curves are highlighted with a D and an S, respectively. The curves cross at the fixed price p^* and quantity q^*.

equilibrium price is *stable* or *unstable.*[7] The solution to the difference equation (2.5) (see Appendix A.1 for the derivation) is given by

$$p_t = \left(\frac{s_1}{d_1}\right)^t \left(p_0 - p^*\right) + p^*. \tag{2.9}$$

The solution (2.9) shows that the necessary condition for the price p_t, starting from an initial condition p_0, to converge to the equilibrium value p^* is that the absolute value of the ratio $\frac{s_1}{d_1}$ is less than 1. As we can see from (2.9), when this condition holds and as $t \to \infty$, the price p_t converges to p^*. Note that the ratio is nothing else but the relative slopes of the supply and demand curves, see (2.1) and (2.2).

The next question is to study the nature of convergence towards the stable equilibrium, i.e. is it going to be an oscillatory convergence or a monotonic convergence? The simple answer is that if $\frac{s_1}{d_1}$ is negative we have oscillations and if $\frac{s_1}{d_1}$ is positive we have monotonic dynamics. The stability condition and the analytical condition for the nature of convergence (or divergence) are summarised in Table 2.1.

In Table 2.1 the stability conditions are given along the columns and the conditions for the nature of convergence (or the *disequilibrium dynamics*) are read along the rows. For instance, starting from any initial price, the condition for the price to converge to the equilibrium value p^* is given by $|s_1/d_1| < 1$ (column 1). At the same time, if $s_1/d_1 < 0$ (row 1) the disequilibrium dynamics exhibits oscillations in the price *close* to p^*.[8]

On the other hand, if $s_1/d_1 > 0$ (row 2) the disequilibrium dynamics exhibits monotonic convergence to the equilibrium price p^*. Similarly, the system becomes unstable when $|s_1/d_1| > 1$ (column 2) and for the disequilibrium price dynamics to become oscillatory and monotonic the conditions can be read off from rows 1 and 2 respectively 2 of Table 2.1.

Table 2.1: Stability and behavioural conditions

Condition	$\left\|\frac{s_1}{d_1}\right\| < 1$	$\left\|\frac{s_1}{d_1}\right\| > 1$
$\frac{s_1}{d_1} < 0$	stable and oscillatory	unstable and oscillatory
$\frac{s_1}{d_1} > 0$	stable and monotonic	unstable and monotonic

We can also check this numerically from our example in Figure 2.3. In Figure 2.3(a), for the stable and oscillatory case, the equilibrium price is

$$p^* = (s_0 - d_0)/(d_1 - s_1) = (0.5 - 4)/(-1.7 - 1.1) = 1.25$$

with $s_1/d_1 = 1.1/-1.7 \approx -0.65$ and in Figure 2.3(b), the oscillatory divergence case, the equilibrium price

$$p^* = (s_0 - d_0)/(d_1 - s_1) = (0.5 - 4)/(-1.7 - 1.9) \approx 0.97$$

with $s_1/d_1 = 1.9/-1.7 \approx -1.12$, which both follow the results in Table 2.1

2.3.1 COBWEB PLOTS

In this section, we provide an alternative way to visualise the dynamics of the model through – so-called *cobweb plots.* This is a popular representation of the cobweb model in the undergraduate textbooks and highlights the disequilibrium dynamics in an alternative way, but one that is still based on the equations and graphical representations introduced in Figure 2.2.

First we rewrite equations (2.1) and (2.2) using the general notation for quantity q instead of D and S and ignoring the time subscripts, for now, and get

$$q = d_0 + d_1 p, \quad d_0 \geq 0, d_1 < 0, \tag{2.10}$$

$$q = s_0 + s_1 p, \quad s_0 \geq 0, s_1 > 0, \tag{2.11}$$

which are two separate dependencies (demand and supply) between price and quantity. We can also swap the order of the dependencies between p and q so that

$$p = -\frac{d_0}{d_1} + \frac{1}{d_1} q, \quad d_0 \geq 0, d_1 < 0, \tag{2.12}$$

$$p = -\frac{s_0}{s_1} + \frac{1}{s_1} q, \quad s_0 \geq 0, s_1 > 0. \tag{2.13}$$

The two functions (2.12) and (2.13) (or equivalently (2.10) and (2.11)) are plotted in Figure 2.4, for the same parameter values as in Figure 2.3. Since we are interested in how (q, p) evolves over time we consider (2.10) and (2.13) and reintroduce time to get

$$q_t = d_0 + d_1 p_{t-1}, \quad d_0 \geq 0, d_1 < 0, \tag{2.14}$$

$$p_t = -\frac{s_0}{s_1} + \frac{1}{s_1} q_t, \quad s_0 \geq 0, s_1 > 0, \tag{2.15}$$

where the former equation corresponds to the demand function and the latter to the supply function.

We can now determine how the quantity q_t and price p_t evolve at every step of the iteration using the following steps:

The cobweb algorithm

1. Given the price p_{t-1} at time t, q_{t-1} is the quantity that the consumer wants to buy at time t as given by (2.14).
2. Given that the consumer wants to buy quantity q_t at time t, the seller offers price p_t at time t as given by (2.15).
3. Go back to step 1.

This process can be seen as a haggling between the seller and buyer until a common price can be agreed upon. In Figure 2.5, we can see how the

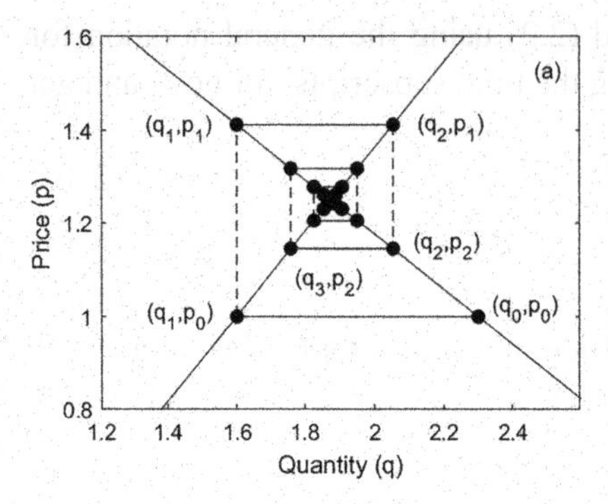

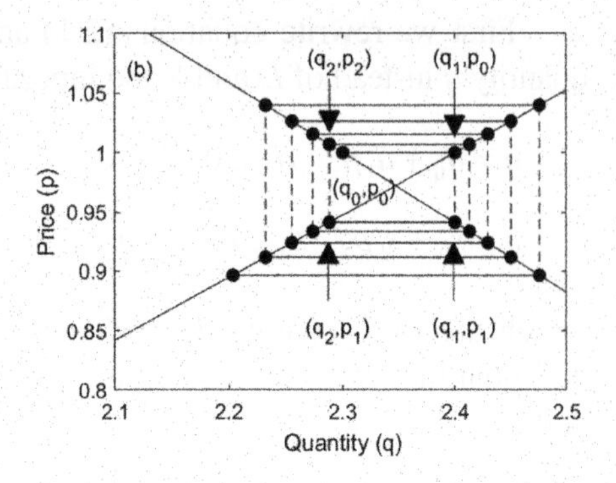

Figure 2.5: Evolution of price p_t and quantity q_t over time t corresponding to Fig. 2.3. Panel (a) shows the stable case, where the market clears, and the (b) panel shows the unstable case, where the market does not clear. The respective parameter values are $s_1 = 1.1$ in (a) and $s_1 = 1.9$ in (b).

haggling process unfolds at each time step starting from an arbitrary initial price and quantity pair (q_0, p_0). Figure 2.5 (a) shows the convergence of the haggling process whereas Figure 2.5 (b) shows the divergence or breaking down of haggling between the buyer and the seller. The two cases in Figure 2.5 correspond to the cases in Figure 2.3 (a) and (b), respectively, and visualise the unfolding of the haggling process at each time step that underpins the convergent and divergent price dynamics.

Together with the MATLAB® codes for simulating and generating the time histories in Figure 2.3, introduced earlier, the MATLAB® code to generate the analysis plots in Figures 2.4 and 2.5 is given below.

Visualisation

```
figure(2),hold on,
plot( [0,d0],[-d0/d1,0],'k',[s0,d0],[0,(d0-s0)/s1],'k')
plot((d1*s0-d0*s1)/(d1-s1),(s0-d0)/(d1-s1),'ko',...
    'MarkerSize',6,'MarkerFaceColor','k',...
    'MarkeredgeColor','k')
box on, hold off
ylabel('Price (p)')
xlabel('Quantity (q)')

figure(3),hold on,
D = d0 + d1*p; %Demand
S = s0 + s1*p; %Supply
```

```
plot([0,d0],[-d0/d1,0],'k',[s0,d0],[0,(d0-s0)/s1],'k')
for k=1:n
    plot([D(k) S(k)],[p(k) p(k)],'k-o',...
        'MarkerFaceColor','k')
    if mod(k,2)==0 && k>1
        plot([S(k-1),D(k)],[p(k-1),p(k)],'k--')
    elseif k>1
        plot([D(k),S(k-1)],[p(k-1),p(k)],'k--')
    end
end
box on, hold off
axis([2.1,2.5,0.8,1.1])
%axis([1.2,2.6,0.8,1.6]) %Figure 2.3(b)
ylabel('Price (p)')
xlabel('Quantity (q)')
```

2.4 CONCLUDING REMARKS, EXTENSIONS AND CHALLENGES

The cobweb model discussed in this chapter provides a basic analytical framework to study the dynamics at and around the demand-supply equilibrium in the microeconomic setting. The model, set in discrete time, provides a basis for understanding the underlying disequilibrium dynamics of a market in terms of monotonic and oscillating behaviours. The conventional textbook treatment of the model uses comparative statics to study the impact of demand or supply shocks to the model, i.e. by shifting the demand and supply curves due to various shocks and *comparing* various equilibrium positions, without explaining the story of how the system moves from one equilibrium to another. Such an analysis assumes that the stability of the equilibrium price is always guaranteed and neglects its context and the out-of-equilibrium dynamics, which is clearly very important for applied policy purposes. As we discussed above, the market does not clear when the equilibrium is unstable. Therefore instability of markets is clearly a cause of concern for both the policy makers and the market participants as it may lead to wider economic crisis. Even persistent oscillations about a stable equilibrium price, like the never-ending negotiations case in the box below, would introduce additional costs for both buyers and sellers from the market efficiency point of view, so policy measures to quickly dampen the oscillations is always preferred.

Task: Never-ending negotiations

There is a case where the model gets *stuck* in a stable oscillation mode with a constant amplitude about the equilibrium point – akin to negotiations going round in circles! Experiment with the simulation parameters to find under what conditions the model generates oscillations with constant amplitude. You can also find the parameter values analytically.

The model can also be generalised by incorporating future *expectation* behaviour. Let us consider the following generalisation

$$D_t = d_0 + d_1 p_t, \qquad d_0 \geq 0,\ d_1 < 0,$$

$$S_t = s_0 + s_1 p_t^e, \qquad s_0 \geq 0,\ s_1 > 0,$$

where the supply is a function of *expected price* p_t^e, rather than the price in the past time. Since $s_1 > 0$ the slope of the supply function is positive, which means that the supply of the produce is positively related to the expected price, i.e. if the supplier expects a higher price for the produce at the next time period $t + 1$, the supply will be increased at time t.

In order to complete the model, we need to specify how the suppliers would determine the expected price, i.e. how suppliers form expectations about the future price p_t^e. This is one of the most challenging problems in economics, and in fact it is this behavioural aspect that distinguishes economics from other scientific disciplines. The rules of forming expectations about the future vary across individuals and contexts. Therefore, articulating a general theory of expectations is a very difficult challenge in economics; nonetheless, we will see in the next chapter how the basic cobweb model can be extended to study market dynamics under expectations.

NOTES

1 There are alternative heterodox approaches in microeconomics, like the post-Keynesian approach, where the decision-making process is viewed through the prism of asymmetric power, for instance between sellers and buyers, firms (owners) and households (workers), or shareholders and managers etc. See Lavoie [1] for a lucid elucidation of post-Keynesian microeconomics.

2 The Dynamical Systems theory uses the term *fixed point*, which is clearer than the usage of the term *equilibrium* in Economics.
3 It is also possible to study the relationship between demand and supply in *continuous* time. However, since most of the economic data are recorded at discrete intervals of times, the discrete-time models are more intuitive. For example, the GDP data is reported quarterly in most countries. Therefore, empirical policy analysis stipulates the use of discrete-time models.
4 Note that we provide a brief explanation of the commands in this chapter. However, from the next chapter onwards, except in some instances, we will include self-explanatory comments as preamble to the codes. Note also that all commands that follow "%" in the MATLAB® environment are treated as comments and are not evaluated.
5 See Appendix A.1 for the general treatment of the mathematical methods for solving difference equations.
6 For the price p^* to be positive the inequalities $d_0 > s_0 \geq 0$ have to hold since $d_1 < 0$ and $s_1 > 0$ (see (2.1)–(2.2)).
7 In general, when an equilibrium p^* is stable it is often referred to as an "attractor" and when it is unstable it is referred to as a "repellor".
8 From the topological point of view, the word 'close' means infinitesimally near the equilibrium point, i.e. we are only concerned with *local* stability of the equilibrium, not *global* stability. However, since the demand-and-supply model is linear the local stability condition applies globally.

BIBLIOGRAPHY

[1] Lavoie, M. (1994), *Foundations of post-Keynesian economic analysis*, New Directions in Modern Economics, Edward Elgar Publishing.
[2] Herbert, S. (1986), Rationality in Psychology and Economics, In Hogarth R.M. and Reder M.W. (eds.) *Rational Choice: the Contrast Between Economics and Psychology*, University of Chicago Press, pp. 25–40.
[3] North, D.C. (1995), Economic Theory in a Dynamic Economic World, *Business Economics*, Vol. 30, No. 1, pp. 7–12.
[4] Kaldor, N. (1934), A Classificatory Note on the Determination of Equilibrium, *Review of Economic Studies*, Vol. 1, pp. 122–136.

3 EXPECTATION DYNAMICS IN THE COBWEB MODEL

3.1 ECONOMIC PROBLEM

There is one significant difference between physical and economic systems – in the former, only past and current states matter, while in the latter, expectations about the *future* (alongside past and current states) play a part in determining current actions. The economic system is made up of a multitude of economic actors such as individuals, firms, institutions, governments etc., and their economic decisions are driven by their **expectations** about the future. In that sense, the future influences agents' current decisions and will thus affect the current market dynamics. Financial markets are the most obvious example highlighting the crucial role expectations play in determining the current prices on a variety of financial assets (currency, stocks, bonds etc.), where an individual agent's decision to buy, sell or hold a financial asset is mainly driven by the expected future price of the asset. The agricultural commodity markets provide another example of how expectations play a role in (re-)allocating resources to the current production.

The role of expectations in explaining market volatility, particularly in financial markets that are driven by "news" about economic fundamentals, has always been debated, both in academic literature and in the media. In the economic literature, the self-referential process of future expectations affecting current realisations of variables, which in turn lead to a revision of the expectations, underpins many of the models of inflation, exchange-rate volatility and financial market volatility models. Therefore, understanding the role of expectations for the stability of markets is important for both practical investment decisions at the individual level as well as macroeconomic policy considerations. In particular, specifying how agents form expectations about the future becomes crucial for determining the dynamics and stability of the market equilibrium.

Since the introduction of adaptive expectations by Muth [1], the *rational expectations hypothesis* (REH) proposed by Lucas [2] has become the dominant paradigm in economics and finance. According to the REH,

DOI: 10.4324/9780429324406-3

agents' subjective expectation about the future, taking into account all available information, coincides with the mathematical expectation conditional on the same information set. In simple terms, agents' subjective forecasts, based on their knowledge of the "correct" model of the economy, coincide, on an average, with realised outcomes. If this is indeed true, how does the REH explain observed market fluctuations in price and other variables? The REH hypothesis, in the general form, argues that the fluctuations are due to forecasting errors of rational agents, due to unanticipated shocks, but it assumes that they will *learn* from these errors and revise their expectations in such a way that the average forecast error tends to zero. In a sense, the REH argues that the fluctuations observed in economic data are due to rational response of agents to various shocks. Thus, it is argued that market dynamics, under rational expectations, becomes more stable than other forms of expectation behaviour by agents.

The REH imposes a heavy burden on agents' ability to *know* the "correct" or the mathematical model of the economy. In recent decades, attempts have been made to generalise the rationality assumption via the notion of *bounded rationality* [3]. In a bounded rational-agent economy, the agents use their subjective or *perceived* model of the economy to *learn* the correct model of the economy from the past data using econometric techniques [4]. This induces a relationship, or a mapping, between the perceived and the correct model of the economy. The rational expectations equilibrium is a fixed point or the equilibrium point of that mapping and the bounded rational agents learn how to converge towards it. The literature on this topic discusses how different types of learning such as *ordinary least squares learning* [5], *genetic algorithm learning* [6] and other types of adaptive learning techniques enforce convergence to the *rational expectations equilibrium* (REE). It argues that if convergence occurs it implies that the REE is an accurate description of the realised market equilibrium outcome, see Hommes *et al.* [7]. However, there are counter results showing how adaptive-learning techniques may not lead to convergence to the REE. For instance, Hommes [8] uses the cobweb model to show how adaptive learning need not lead to convergence to the REE. Other articles, such as Grandmont and Laroque [9], show how learning may even lead to instability.

With this somewhat simplified introduction to a highly technical and voluminous literature as a background context, let us proceed to the chapter with a modest objective of studying the influence of expectations on the dynamics and the stability properties using a simple demand-supply model. For the purposes of continuity and for pedagogical reasons, we will use a *generalised cobweb model* and study the role of expectations on the stability properties of the model.

3.2 MODELLING

To model price expectation in the demand and supply model we will use a generalised cobweb model, where the supply decisions are based on *expected price,* rather than past price. In such case, the basic demand and supply model (2.1)-(2.2) is generalised as

$$D_t = d_0 + d_1 p_t, \quad d_0 \geq 0, d_1 < 0 \tag{3.1}$$

$$S_t = s_0 + s_1 p_t^e, \quad s_0 \geq 0, s_1 > 0 \tag{3.2}$$

for $t = 1,2,3, \ldots,$ where, as in Chapter 2, D_t represents demand at time t and is negatively related ($d_1 < 0$) to the price p_t at time t. However, supply (S_t) at time t is now a function of the expected price (p_t^e) at time t, rather than the past price, and is positively related ($s_1 > 0$) to it.[1]

In order to complete the model, we need to articulate specific rules or models of expectation formation by the suppliers. For the purposes of illustration, we use two simple models of expectations and study the dynamics of the generalised cobweb model (3.1)-(3.2).

Naïve expectation. First, let us assume that the supplier is *naïve* and expects that the next period's price will be the same as the previous period's price, so that

$$p_t^e = p_{t-1} \tag{3.3}$$

Task1: Reduce the generalised cobweb model under the naïve expectation rule.

Show that the naïve-expectation rule reduces the generalised cobweb model (3.1)-(3.2) to the original cobweb model (2.1)-(2.2) described previously in Chapter 2.

Normal price expectation. Second, we assume that the supplier has a notion of a *normal price* (or a *long-run price*) p_N towards which the current market price will tend over time. A simple way to formalise this expectations rule is by letting

$$p_t^e = p_{t-1} + c(p_N - p_{t-1}), \quad 0 < c < 1, \tag{3.4}$$

where the parameter c is referred to as the *speed of adjustment*. The intuition behind this rule is that if the price in the previous time $t-1$ is lower than the normal price p_N, then the supplier would expect a higher price in the current time period relative to the previous time period's price. The parameter c reflects the speed with which the previous time period's price adjusts towards the normal price. Note that when $c = 0$ the expected price $p_t^e = p_{t-1}$, yields the naive-expectation case, and when $c = 1$ the expected price $p_t^e = P_N$, yields the instantaneous adjustment to the normal price. Also, when c is close to 0 (or 1), the supplier expects the convergence to the normal price to be slower (or faster). The choice of normal price p_N can, of course, be argued as it requires some notional knowledge of the system. To keep things simple, we will choose the normal price as the equilibrium price (2.7) of the original demand and supply model (2.1)-(2.2) such that

$$p_N = \frac{s_0 - d_0}{d_1 - s_1}. \tag{3.5}$$

The market clearing condition is the same as in Chapter 2, namely,

$$D_t = S_t, \tag{3.6}$$

which together with (3.1)-(3.2) is the generalised cobweb model where expectation rules are introduced, for instance (3.3) or (3.4). Let us next analyse the stability properties of the system based on these expectation rules.

3.3 ANALYSIS, SIMULATION AND VISUALISATION

Since the naïve expectation model (3.3) results in the original cobweb model discussed in Chapter 2, we focus our attention on the normal price expectation model (3.4). First, substitute equations (3.1) and (3.2) into the market clearing condition (3.6) that gives

$$p_t = \frac{s_1}{d_1} p_t^e + \frac{s_0 - d_0}{d_1}. \tag{3.7}$$

Next, using the normal price expectation rule and substituting (3.4) into (3.7) yields

$$p_t = \frac{s_1}{d_1}(1-c)p_{t-1} + \frac{s_1 c p_N + s_0 - d_0}{d_1}, \tag{3.8}$$

a first-order difference equation that describes the evolution of the price p_t for the generalised cobweb model with the normal-price expectation rule, similar to (2.5) in Chapter 2.

To get an intuition of the behaviour of the demand and supply model with expectation, let us start by simulating the difference equation (3.8) using the same parameter values and initial condition as in Section 2.3, see the box below.

Parameter values and initial conditions used in simulations

The same parameter values and initial condition are used here as in Section 2.3, namely,

$$d_0 = 4,\ d_1 = -1.7,\ s_0 = 0.5,\ s_1 = 1.1 \text{ and } p_0 = 1.$$

We also need to set a value for the speed of adjustment, and to start with let us set the value as

$$c = 0.8.$$

Recall that these values are chosen arbitrarily for the purpose of illustration. You are encouraged to test different parameter values (even non-sensible ones) in order to get a better feeling for dynamics of the system.

The MATLAB® code for simulating and visualising the output is very similar to what we had in Section 2.3, and is given below. The main difference is the addition of the speed of adjustment parameter c.

Demand and supply with normal-price expectation rule

```
%Demand and supply with normal price expectation rule
%Preamble
clear all % Clears all variables
close all % Closes all figure windows

%Parameters
d0 = 4;
d1 = -1.7;
s0 = 0.5;
s1 = 1.9;
c  = 0.8;

%Initial Condition
p0 = 1;
```

```
%Normal price
pN = (s0-d0)/(d1-s1);

%Simulation parameters
n=10; % Number of iterations

%Simulation
p(1) = p0;
for k = 1:n
   p(k+1) = (s1/d1)*(1-c)*p(k) + (s1*c*pN+s0-d0)/d1;
end

%Visualize time versus price
figure(1)
plot(0:n,p,'ko','MarkerFaceColor', 'k')
xlabel('t'),ylabel('p_t')
```

Alternatively, if we want to use the original formulation of p_t^e (see equation (3.4)) for the expected price, we can instead use the following `for`-loop for simulation.

Simulation with price expectation

```
%Simulation
p(1) = p0;
pk(1) = p0;
for k = 1:n
   pe(k+1) = p(k) + c*(p_N - p(k));
   p(k+1) = (s1/d1)*pe(k+1) + (s0-d0)/d1;
end
```

In Figure 3.1 we analyse two scenarios for the demand-and-supply model with normal price expectations, where the speed of adjustment parameter c plays an important role in capturing suppliers' expectations. We analyse the effect of the speed of adjustment parameter on the stability of the equilibrium, we use the same initial price p_0 but with two different values of parameter c. Figure 3.1(a) shows the case with the speed of adjustment parameter set at $c = 0.8$, as introduced in the box above, and in Figure 3.1(b) the value is set at $c = 0.2$. The MATLAB® code to simulate the modified system is shown below.

Parameter values

```
%Parameters
d0 = 4;
d1 = -1.7;
s0 = 0.5;
s1 = 1.9;
c = 0.2;
```

Comparing the two cases in Figure 3.1, we can see that starting from the same initial condition the market converges to the equilibrium price faster in Figure 3.1(a), where the value of c is higher ($c = 0.8$) than in Figure 3.1(b), where the value of $c = 0.2$. Therefore, from the simulation results, it is clear that the introduction of the normal price expectations seems to impact the speed with which the marker reaches its stable equilibrium. Let us next explore the reason analytically.

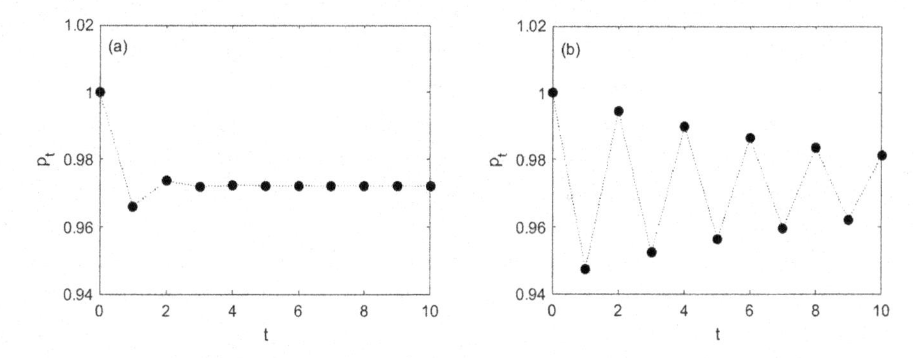

Figure 3.1: Evolution of price p_t over time t in the demand-and-supply model with normal price expectation, where in (a) $c = 0.8$ and in (b) $c = 0.2$. In both cases the initial price $p_0 = 1$.

Question

How does the dynamics of the demand-and-supply model with normal price expectation differ from the original model in Chapter 2? Has the introduction of the expectation rule made the model more (or less) stable?

To answer this question we follow the method used in Chapter 2, where we start by finding the equilibrium price p^* for which the market clearing condition (3.6) holds. However, we will take a slightly different approach than in Section 2.3 in that we use (3.8) directly and determine the location of the the fixed or equilibrium price by letting $p^* = p_t = p_{t-1}$, which yields

$$p^* = \frac{s_1}{d_1}(1-c)p^* + \frac{s_1 c p_N + s_0 - d_0}{d_1} \tag{3.9}$$

and thus

$$p^* = \frac{s_1 c p_N + s_0 - d_0}{d_1 - s_1(1-c)}. \tag{3.10}$$

Now substituting for p_N from (3.5) we get

$$p^* = \frac{s_0 - d_0}{d_1 - s_1}, \tag{3.11}$$

which we will also get if we let $c = 0$ (the naïve expectation rule). Note that we assumed $p_N = p^*$, which implies that the suppliers' expectations do not influence the equilibrium price of the market and the market discovers the equilibrium/normal price. However, in the case when agents adapt their expectations about the equilibrium price, a self-reinforcing dynamic between agents' expectations and market outcome, it can lead to multiple equilibria. In such cases, depending on agents' adaptation behaviour, the market could exhibit diverse dynamics, including the boom-bust dynamics as regularly exhibited in financial market data.

It is clear from the above simulation analysis that suppliers' expectations influence the stability properties of the market equilibrium. To further understand how the introduction of expectation impacts the stability properties, let us solve the difference equation (3.8). The solution to the difference equation (3.8) is given by (see Appendix A.1)

$$p_t = \left(\frac{s_1}{d_1}(1-c)\right)^t \left(p_0 - p^*\right) + p^* \tag{3.12}$$

or equivalently

$$p_t = \left(\frac{s_1}{d_1}(1-c)\right)^t \left(p_0 - \frac{s_0 - d_0}{d_1 - s_1}\right) + \frac{s_0 - d_0}{d_1 - s_1}. \tag{3.13}$$

The solution (3.13) shows that the necessary condition for price p_t, starting from the initial condition p_0, to converge to the equilibrium value p^* is

$$\left|\frac{s_1}{d_1}(1-c)\right| < 1.$$

Further, since $0 < c < 1$, the nature of the disequilibrium dynamics, whether oscillatory or monotonic, is determined by

$$\frac{s_1}{d_1} < 0 \quad \text{or} \quad \frac{s_1}{d_1} > 0,$$

respectively. Thus, as in the original cobweb model in Chapter 2, the disequilibrium dynamics simply depends on the relative slopes of the supply and demand curves, the parameters s_1 and d_1 of the model.

The stability condition in this generalised model is similar to the original cobweb model in Chapter 2, except the multiplicative scaling term $1-c$. The parameter c reflects the strength of suppliers' expectations regarding the actual price converging to the normal price. When c is higher, it implies that the suppliers expect the convergence to the "normal" price is faster, and therefore the market clears more quickly, as seen in Figure 3.1(a). On the other hand, when suppliers expect the convergence to the normal price to be slow, the market takes a longer time reach its equilibrium price, as shown in Figure 3.1(b), where the value of c is smaller. All these conditions are summarised in Table 3.1.

Table 3.1 Stability and behavioural conditions

Condition	$\left\|\frac{s_1}{d_1}(1-c)\right\| < 1$	$\left\|\frac{s_1}{d_1}(1-c)\right\| > 1$
$\frac{s_1}{d_1} < 0$	stable and oscillatory	unstable and oscillatory
$\frac{s_1}{d_1} > 0$	stable and monotonic	unstable and monotonic

We can also check this numerically from our example in Figure 3.1. In both Figure 3.1(a) and (b), the value of the equilibrium price is

$$p^* = \frac{s_0 - d_0}{d_1 - s_1} = \frac{0.5-4}{-1.7-1.1} = 1.25.$$

However, as mentioned above, in Figure 3.1(a) $c = 0.2$ and in Figure 3.1(b) $c = 0.8$. The numerical value of the stability condition is

$$\frac{s_1}{d_1}(1-c) = \frac{1.9}{-1.7}\cdot 0.2 \approx -0.22$$

in Figure 3.1(a) and

$$\frac{s_1}{d_1}(1-c) = \frac{1.9}{-1.7}\cdot 0.8 \approx -0.89$$

in Figure 3.1(b). This is interesting since, as we saw in Chapter 2, for the same parameter values ($s_1 = 1.9$ and $d_1 = -1.7$) the system was unstable, but now with the introduction of the expectations parameter c, we see that the system has turned stable.

Moreover, we can calculate for the range of parameter values c for which the equilibrium price is stable. From (2.9) we have

$$\left|\frac{s_1}{d_1}(1-c)\right| < 1 \quad \Leftrightarrow \quad 1+\frac{d_1}{s_1} < c < 1-\frac{d_1}{s_1}.$$

For example, the numerical values of the parameters used in Figure 3.1 yield the interval

$$1+\frac{-1.7}{1.9} < c < 1-\frac{-1.7}{1.9} \Leftrightarrow 0.105 < c < 1.895.$$

Of course, this interval is specific to this particular example, but the principle is generic. Since $0 < c < 1$ we see that the interval $0.105 < c < 1$ yields a stable equilibrium price p^* and also validates Figure 3.1.

3.3.1 COBWEB PLOTS

Finally, as we did in Chapter 2, we can visualise the dynamics of the model through cobweb plots, as shown in Figure 3.2. Figures 3.2(a) and (b) respectively correspond to the two cases presented in Figures 3.1(a) and (b). The main difference between the cobweb plots here and the ones of the original model in Chapter 2 is the speed of convergence to the equilibrium price, which depends on the value of the parameter c, as shown in Figure 3.2 and discussed above. Figure 3.2(a), with a higher value of c, implies that the suppliers expect that any deviation from the normal price will vanish quickly and Figure 3.2(b), with a lower value of c, indicates that the suppliers expect a slower convergence to the normal price. Note specially the difference in the cobweb plots in Figure 3.2, *vis-à-vis* the cobweb plots in Figure 2.5 for the

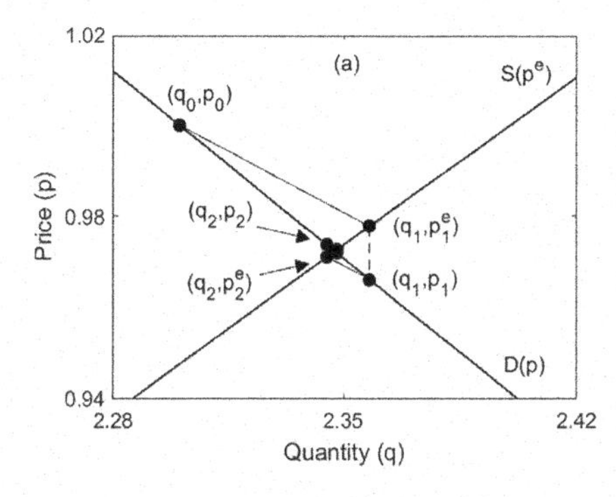

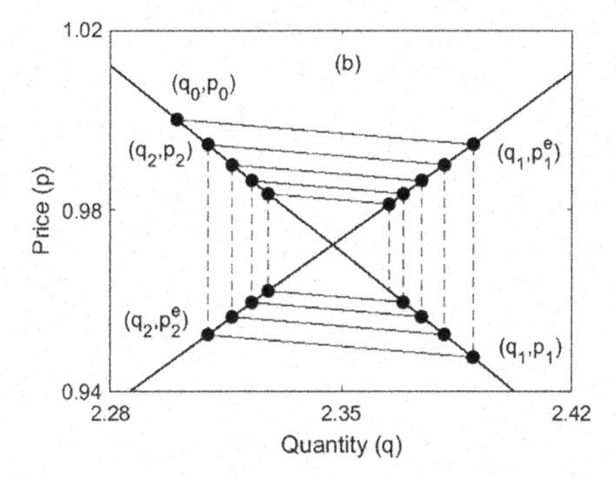

Figure 3.2: Evolution of price p_t and quantity q_t over time t corresponding to Figure 3.1. In (a) $c = 0.8$ and in (b) $c = 0.2$.

original cobweb model in Chapter 2. Here we see a "slant" in the cobwebs, which signifies faster convergence to the equilibrium. This is because the supply function $S(p)$ here depends on the expected price p_t^e, instead of the past period's price p_{t-1}. In the current model, with normal price expectations, the size of the expectation parameter c determines the speed of convergence to the equilibrium. In other words, when suppliers strongly expect (i.e. when $c = 0.8$) that price will converge to the normal price the market reaches its equilibrium quickly, as shown in Figure 3.2(a), and when the suppliers' expectation is weaker (i.e. when $c = 0.2$) the market takes more time to reach its equilibrium value, as shown in Figure 3.2(b). In general, when suppliers, or agents, form expectations about the future, they take into account all available information, and depending on the assumption about the expectation rules, the model suggests the speed with which markets reach their equilibrium values. This can be seen in Figure 3.2, where the slant is more in (a) because the value of the supplier's expectations or learning parameter is higher ($c = 0.8$) than in (b) ($c = 0.2$).

To generate Figure 3.2, a very similar MATLAB® code that generated Figure 2.5 in Chapter 2 is used and given below.

Cobweb plots

```
figure(3),hold on,
D = d0 + d1*p; %Demand
S = s0 + s1*pe; %Supply

plot([0,d0],[-d0/d1,0],'k',[s0,d0],[0,(d0-s0)/si],'k')
for k=2:n
```

```
    plot([D(k-1) S(k)],[p(k-1)pe(k)],'k-o',...
          'MarkerFaceColor','k')
    if mod(k,2)==0 && k>1
       plot([S(k),D(k)],[pe(k),p(k)],'k-- ')
    elseif k>1
       plot([D(k),S(k)],[p(k),pe(k)],'k--')
    end
end

box on, hold off
axis([2.28 2.38 0.95 1.01]) %Figure 3.2(a)
%axis([2.28 2.42 0.94 1.01]) %Figure 3.2(b)
ylabel('Price (p)')
xlabel('Quantity (q)')
```

3.4 CONCLUDING REMARKS, EXTENSIONS AND CHALLENGES

In this chapter, we generalised the original cobweb model in Chapter 2 and investigated the role of expectations in determining the length of time the market could spend in the disequilibrium state before converging to the equilibrium price. In simple terms, we see that expectations influence both whether or not the market will clear and *how long* the market would take to reach its equilibrium. The expectation rules we attributed to suppliers in this model may be considered unrealistic because in real life people adapt their expectations based on the actual realised outcomes. While the rational expectations hypothesis provides the theoretical limit, the *adaptive expectations hypothesis* yields a more realistic alternative in which the expected price is gradually adjusted to the forecast error. More formally, the adaptive expectations rule can be written as

$$p_t^e = p_{t-1}^e + \beta\left(p_{t-1} - p_{t-1}^e\right), \quad 0 < \beta < 1, \tag{3.14}$$

where p_{t-1}^e is the expected price at time $t - 1$, and β is the *adaptation* parameter.

Setting $\beta = 1$, the expectation model (3.14) reduces the generalised cobweb model to the original cobweb model in Chapter 2. The economic intuition here is if the observed value in the previous period is greater, i.e. $p_{t-1} > p_{t-1}^e$ (or smaller, i.e., $p_{t-1} < p_{t-1}^e$) then the agent will revise the current expected value p_t^e upwards (or downwards).

Task 2: Solving the model under adaptive expectations

Solve model (3.1)-(3.2) using the adaptive expectations rule (3.14) and verify whether you get the solution

$$p_t = \left(p_0 - p^*\right)\left(\beta\left(\frac{s_1}{d_1} - 1\right)\right)^t + p^*$$

or not and determine the stability condition.

Task 3: Compare the stability conditions

Comparing the stability condition of the original model, given by

$$-1 < \frac{s_1}{d_1} < 1,$$

with the result in Task 2 and discuss how the introduction of the adaptive expectations changes the stability of the cobweb model?

Hint: Use numerical values for β and determine the range of values of β that leads to a stable equilibrium price.

Task 4: Simulate and plot the adaptive expectations model

Simulate and plot how the price varies in the adaptive expectations case vis-a-vis the expectation models introduced in this chapter. Experiment with different parameter values and initial prices.

Hint: Adapt the codes used in Chapter 2 and in this chapter.

NOTE

1 To clarify the notation, it denotes the price that is expected to materialise at time t. A more elaborate expression is $E_{t-1}[p_t]$, i.e., expectation made at time t–1 regarding the price at time t. For notational simplicity, we will use the notation p_t^e to denote the expected price at time t.

BIBLIOGRAPHY

[1] Muth, J.F. (1961), Rational Expectations and the Theory of Price Movements, *Econometrica,* Vol. 29, pp. 315–335.

[2] Lucas, R.E. (1972), Econometric Testing of the Natural Rate Hypothesis, In Eckstein O. (ed.), *The Econometrics of Price Determination Conference,* pp. 50–59.

[3] Sargent, T.J. (1993), *Bounded Rationality in Macroeconomics,* Oxford: Clarendon Press.

[4] Evans, G.W. and Honkapohja, S. (2001), *Learning and Expectations in Macroeconomics,* Princeton, NJ: Princeton University Press.

[5] Bray M.M. and Savin N.E. (1986), Rational Expectations Equilibria, Learning, and Model Specification, *Econometrica,* Vol. 54, issue 5, pp. 1129–1160.

[6] Arifovic, J. (1994), Genetic Algorithm Learning and the Cobweb Model, *Journal of Economic Dynamics and Control*, Vol. 18, pp. 3–28.

[7] Hommes C.H., Sonnemans J., Tunistra J. and Van De Velden, H. (2007), Learning in Cobweb Experiments, *Macroeconomic Dynamics,* Vol. 11(1), pp. 8–33.

[8] Hommes, C.H. (1994), Dynamics of thc Cobweb Model with Adaptive Expectations and Nonlinear Supply and Demand, *Journal of Economic Behavior and Organization*, Vol. 24, pp. 315–335.

[9] Grandmont, J.M. and Laroque, G. (1991), Economic Dynamics with Learning: Some Instability Examples, In Barnett W.A., Cornet B., d'Aspremont C., Gabszewicz J. and Mas-Colell A. (eds.), *Equilibrium Theory and Applications,* Cambridge: Cambridge University Press, pp. 247–273.

4 KEYNESIAN MULTIPLIER MODEL

4.1 ECONOMIC PROBLEM

In the previous chapters we studied the dynamics of a single market in a microeconomic setting. We now move on to analyse the *macroeconomy,* which is the study of the whole economy encompassing all the markets, such as labour market, product market, financial market and all the institutions such as governments, central banks etc. Alternatively, one could think of the macroeconomy as a network of markets and institutions interconnected through multilevel feedback mechanisms that evolve over time. Therefore, the study of macroeconomics – the study of the whole economy – deals with an evolving *complex system*. There are two methodological approaches in macroeconomics.

The first approach takes the route of explaining the macroeconomy through aggregating individual agent decisions, at the micro level. This, so called, *microfoundations* approach is referred to as the *methodological individualism,* where the macroeconomic outcomes are explained by the subjective preferences and choices of individual agents. In this approach, the macroeconomic level of employment is a consequence of individual level decisions of supply and demand for labour by workers and firms, respectively. Both workers and firms are assumed to be optimising agents, whose decisions of supply of labour and demand for labour are driven purely by their own utility and profit maximizing behaviour respectively. In dynamic models, the inter-temporal optimisation, i.e. the decision to consume today and save for tomorrow, of the *rational agents* drives product, labour and capital markets and determines macroeconomic output and employment in the economy. Prices act as the fundamental mechanism that coordinates demand and supply in these markets and thereby leads to the overall macroeconomic equilibrium of the economy. Fortified by further assumptions such as homogeneity of rational agents in an economy with single (composite) good and other mathematical restrictions on the utility and the production functions, the *existence, uniqueness* and *stability* of the macroeconomic

DOI: 10.4324/9780429324406-4

equilibrium are guaranteed in these models. The family of such models is referred to as the *neoclassical general equilibrium models*.[1]

The second approach to study the macroeconomy is to look at the overall *structure* of the economy using aggregate variables such as consumption, investment, government expenditure and net exports of the economy. In this approach, the dynamics of the macroeconomy is studied through the evolution of the aggregate variables. We will see in the later chapters that macroeconomic growth is studied by looking at how the aggregate variables such as the investment expenditure, i.e., the stock of tangible physical capital or human capital, evolve over time. In that sense, this approach emphasises the *structural* conditions of production, influenced by socio-economic-political factors, in the economy that drives the evolution of aggregate variables such as investment, human capital, etc. In this structural approach, individual agents' behaviour is treated as exogenous parameters such as the marginal propensity to consume or the marginal propensity to import, etc., in contrast with the micro-foundations approach.

The structuralist approach

The term *structuralist macroeconomics* is associated with the work of the Economic Commission for Latin America (ECLAC), also known as CEPAL, and has its roots in the works of John Maynard Keynes [5], Michal Kalecki [4], Joan Robinson [8], Roy Harrod [2], Luigi Pasinetti [7] and their followers. From a methodological point of view, the guiding principle of this approach lies in the idea that the economic system should be studied as an organic system of inter-related constituent elements and not as a collection of individual elements considered in an atomistic manner. In other words, the *relations* between the elements that constitute the economic and social structure, and their evolution, are important as opposed to the a-contextual or a-historical behaviour and actions of individual elements of the system.

The *Keynesian economics* and the schools of thought in this lineage, like the post-Keynesian economics, follow the structural approach. The aggregate output and employment are demand driven in the sense that they are determined by expenditures of households, firms and governments. For instance, the consumption and investment expenditures determine aggregate output and employment. Unlike what we see in the neoclassical general equilibrium approach, the level of investment in the economy is independent of the level of saving. In the general equilibrium models, these two decisions are indistinguishable, so that the total income is divided between consumption and saving (which is nothing but investment), which leads to the Say's law story of saving leading to investment and the determination of output and employment in the economy. In contrast, in the Keynesian economics, this causation is reversed where the aggregate saving is determined by the aggregate level of investment.

Schools of Thought in Macroeconomics

The umbrella term *Keynesian economics* is used to include different schools of thought, such as the older *neo-Keynesian* and the more modern *new Keynesian* schools, all of which propose a theory that provides a *synthesis* that combine both the general equilibrium theory and Keynesian ideas. For example, the older neo-Keynesian approach advanced the IS-LM model's (see Chapter 5) interpretation of Keynes' *General Theory* and the current new Keynesian approach advocates the class of *consensus* models known as the *dynamic stochastic general equilibrium* (DSGE) that incorporates some key Keynesian ideas such as *nominal rigidities* in the representative rational agent general equilibrium framework [1].

With the macroeconomic equilibrium condition being that saving is equal to investment, one of the fundamental differences between the neoclassical general equilibrium and the Keynesian approach is in the *direction of causation* between investment and saving. The neoclassical general equilibrium theory proposes the causation from saving to investment, and the Keynesian approach suggests the opposite direction of causation that is starting from investment to saving. From a policy perspective, the direction of causation between saving and investment underlies the diametrically opposite policy prescriptions arising from these approaches. For instance, when an economy is in recession, the Keynesian approach would suggest an increase in government expenditure to stabilise the markets, whereas the neoclassical general equilibrium approach would suggest a reduction of government expenditure to enable the markets to lead the recovery.

In this and the following chapters, we will study the Keynesian approach for the determination of output and employment, through the so called *multiplier* process and study the stability properties of the macroeconomic equilibrium. In Chapter 12 we will study the *real business cycle* model, which forms the basis for a class of *dynamic stochastic general equilibrium* (DSGE) models.

4.2 MODELLING

The macroeconomic output, or gross domestic product (GDP), is estimated in three different ways using measures such as *expenditure, value of final goods and services*, and *income*.

The *expenditure method* measures GDP as the total expenditures incurred by households, firms, governments etc., on various goods and services so that

$$Y \equiv C + I + G + (X - M), \tag{4.1}$$

where Y denotes output (or GDP), C denotes *consumption expenditure, I* denotes *investment expenditure,* G is the *net government expenditure* (i.e. expenditure (E)−revenue (T)), and $X - M$ denotes *net exports* (exports X−imports M). Note that the government expenditure is the net of transfers (e.g. social security benefits and other subsidies). The expenditure method (4.1) only includes *final* or finished goods. However, the production of final goods and services also uses raw materials and other intermediary goods as inputs, but the expenditure method includes only the purchases of the finished goods.

The *value-added method* addresses the issue of raw materials or inputs by estimating the net value created in all the production sectors and is defined as

$$Y = \text{value of output sold} - \text{cost of raw materials}. \tag{4.2}$$

Finally, in addition to the expenditure and value-added methods we have the *income method.* In this method GDP is calculated as the total income of all agents in the economy, so that

$$Y = W + \Pi, \tag{4.3}$$

where W and Π denote wages or salaries of workers and profits of the owners of capital, respectively.

The three methods of estimating the GDP are identities that must hold at every point in time and, in principle, should yield the same value for GDP. Intuitively, it makes sense that the total income should equal total expenditure because every transaction in the economy has a buyer (spends) and a seller (earns). However, in practice these measures yield different estimates due to measurement errors. In some countries, the income method may be more difficult to estimate due to the, so called, *black economy* and in some cases, the value added measure may yield an underestimation of GDP due to the presence of a large *informal* sector. The expenditure method is relatively less noisy and easier to measure than the other methods, and therefore macroeconomic policy analysis relies mostly on the expenditure method. In the reminder of this chapter we will work with the expenditure method (4.1). The aim here is to study the determination of the equilibrium level of macroeconomic output and then study its stability properties.

Let us consider a closed economy (closed for foreign trade) with government. First we would like to understand the notion of the goods or product market equilibrium and the mechanics of the *multiplier* process. In the Keynesian approach the emphasis is on the demand side and assumes that supply always adjusts to demand, which implies that firms have underutlised production capacity to meet any level of demand.

The standard approach in undergraduate textbooks is to equate the *planned* expenditure with the *actual* expenditure and then describe the multiplier process. In the closed economy, the aggregate demand Y^d is given by

$$Y^d = C + I + G. \tag{4.4}$$

We note that all variables are in real terms or in constant prices. The equilibrium in the goods market is where the planned real expenditure on goods and services (i.e. the aggregate demand) equals the real output, and therefore

$$Y^d = Y. \tag{4.5}$$

The term Y is also income because of the circular nature of the economy, where spending on goods and services becomes income for those producing them.

The next step is to specify the consumption, investment and government expenditure functions. Let us consider a basic, linear consumption function given by

$$C = a + bY, \qquad a \geq 0 \qquad 0 < b < 1, \tag{4.6}$$

where a represents *autonomous consumption,* which is the part of consumption that does not depend on income. But consumption expenditure is driven by income (or disposable income), which is given by the second term of (4.6). The slope parameter b reflects the *marginal propensity to consume* (MPC), the responsiveness of consumption to changes in income.

For simplicity, assuming the investment expenditure $I = \bar{I}$ and government expenditure $G = \bar{G}$ are given constants and substituting these assumptions, and (4.6), in (4.4) yields the aggregate demand for the closed economy

$$Y = a + bY + \bar{I} + \bar{G} \Rightarrow Y = \frac{a + \bar{I} + \bar{G}}{1 - b}. \tag{4.7}$$

We can now use the equilibrium condition (4.5) to study the multiplier process by comparing the planned and actual expenditures in the (Y^d, Y)-plane, which is known as the *Keynesian cross diagram* (similar to Figure 4.1 below), where the planned expenditure is shown on the vertical axis and the actual expenditure on the horizontal axis.

In so far as the multiplier process is concerned, equation (4.7) implies that for a unit increase in either autonomous consumption a, investment expenditure $\bar{I}$ or government expenditure $\bar{G}$, the national income Y will increase by multiplier $\frac{1}{1-b}$ times. With this realisation, we can also use (4.7) for instance, to see how a change ΔI in the investment will lead to a change ΔY in the output by writing

$$\Delta Y = \frac{1}{1-b}(a + \Delta I + \bar{G}). \tag{4.8}$$

The principle of the process can be visualised along the lines as shown in Figure 4.1 in the (Y^d, Y)-plane (planned expenditure – actual expenditure plane).

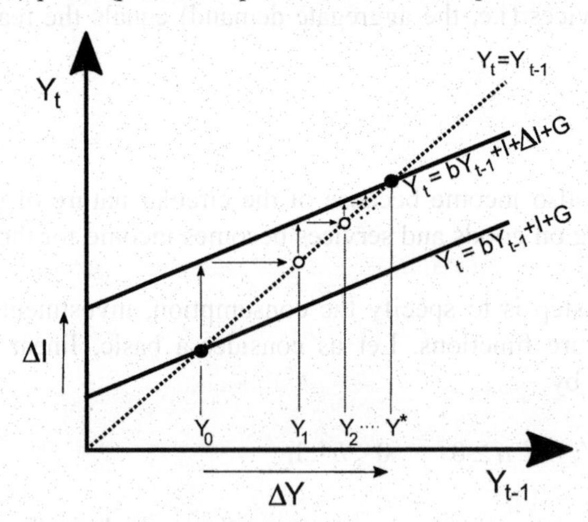

Figure 4.1: The Multiplier process

However, the above presentation can be confusing for novice readers as well as students on two counts. First, the equilibrium condition that the planned expenditure is equal to actual expenditure may confuse readers as the former is exogenously defined rather than arrived at intuitively. Second, explaining the dynamic multiplier process in the static framework in itself can be confusing for some.

A more straightforward way to present the Keynesian multiplier analysis is in its dynamic form. This will alleviate the confusion caused by the static presentation of the multiplier process and allow readers to understand the notion of the equilibrium in a more intuitive way.

Let us restate the national income identity (4.4) in discrete time as

$$Y_t = C_t + I_t + G_t, \tag{4.9}$$

where, as above, Y_t denotes output at time t, C_t denotes consumption expenditure at time t, I_t denotes investment expenditure at time t and G_t is the net government expenditure at time t. Further, let us assume that the consumption expenditure C at time t depends on the income Y at time $t - 1$, so that

$$C_t = a + bY_{t-1}, \tag{4.10}$$

where a is the autonomous expenditure and b the MPC. For simplicity, let us assume that the investment expenditure $I = \bar{I}$ and government expenditure $G = \bar{G}$ are exogenously given constants, and thus

$$I_t = \bar{I} \tag{4.11}$$

and

$$G_t = \bar{G}. \tag{4.12}$$

Substituting assumptions (4.10)-(4.12) to the national income identity (4.9) yields

$$Y_t = bY_{t-1} + a + \bar{I} + \bar{G}. \tag{4.13}$$

Equation (4.13) is a first-order difference equation that provides the basic Keynesian income determination model in a dynamic form. In this formulation, the equilibrium condition is relatively intuitive and is given by the level of income satisfying the condition

$$Y_t = Y_{t-1}. \tag{4.14}$$

This means that the equilibrium level of income is the "fixed" or "stationary" value to which the dynamic income process (4.13) converges. The analytical condition for convergence is given by the stability condition, which we will analyse next in Section 4.3. Intuitively, we can see that the stability of the fixed point is guaranteed under the assumption that $0 < b < 1$, which corresponds to a slope of the output function (4.13) being less then 1 (which is the slope of (4.14)). Note that the MPC parameter determines both the size of the multiplier and the value of the fixed point Y^*. Consider an exogenous increase in the investment expenditure (ΔI), which shifts the actual expenditure line upwards as shown in the Figure 4.1. For a given value of the MPC parameter ($0 < b < 1$), starting from the initial output Y_0, the figure shows the dynamics of adjustment of the system arising from an increase in investment and the convergence to the fixed point Y^*.

Since the equation (4.13) is a first-order difference equation we can analyse this national income model using the same methods as we employed for solving the demand and supply models in Chapters 2 and 3.

4.3 ANALYSIS, SIMULATION AND VISUALISATION

Following the methodology in Appendix A.1, we can find the equilibrium level Y^* of national income Y from the difference equation (4.13) to be

$$Y^* = \frac{a + \bar{I} + \bar{G}}{1 - b} \tag{4.15}$$

and the solution to (4.13) is given by

$$Y_t = (Y_0 - Y^*)b^t + Y^*, \tag{4.16}$$

where Y_0 is the initial value, at $t = 0$, of the national income. Note the similarity with the equilibrium value in (4.15) and the aggregate demand for a closed economy calculated in (4.7).

As was the case for the Cobweb models in Chapters 2 and 3, two questions need to be answered. First, what is the condition for convergence to the equilibrium level of income Y^*? Second, starting from any initial level of income, say Y_0, what is the nature of the movement towards the equilibrium level of income? Can we say whether it will be monotonic or oscillatory? These two questions are important from a policy point of view, as policy makers would like to know if there are any costs associated with the adjustment of the economy to a new level of income.

From both (4.13) and (4.16) we can see that the fixed income is stable if $|b| \leq 1$, which will always be satisfied given the assumption that the economically meaningful range for MPC is $0 < b < 1$. Since the equation (4.13) is linear, we can start from any arbitrary initial level of income (initial condition) Y_0 and the system (economy) will monotonically evolve (since $b \geq 0$) towards the equilibrium level of income Y^*.

The multiplier process

The equilibrium level of income Y^*, given by (4.15), is driven by the so called "multiplier" process. The multiplier is here given by $\frac{1}{1-b}$, where the parameter b is the marginal propensity to consume (MPC), which is a value between 0 and 1. As an example, in economic terms, if $b = 0.6$ it means that individuals consume only 60 per cent of income. Now suppose that there is an increase in any of the autonomous variables in (4.15), for instance, an increase in the exogenous investment expenditure $\bar{I}$. The multiplier process works along the following lines: An initial increase in the investment expenditure increases aggregate demand by ΔI. This increase in output, or income, increases aggregate consumption by $\left(\frac{1}{1-b}\Delta I\right)$, which in turn increases the aggregate demand by the same amount leading to further increase in output and income. This, in turn, leads to a second round increase in aggregate consumption, given by $\left(\frac{1}{1-b}\right)^2 \Delta I$, and so on. The arrows in Figure 4.1 visualise each round of expansion and the eventual convergence towards the equilibrium level of output Y^*.

To simulate and solve the difference equation (4.13) numerically, we can use the following MATLAB® code, which is very similar to the one used in Chapter 2. Note that the entries for the parameter values a, b, $\bar{I}$ and $\bar{G}$, the initial condition Y_0 and the number of iterations n are left empty, so the readers is encouraged to experiment with the system.

Simulation of the Keynesian multiplier model

```
%Preamble
clear all % Clears all variables
close all % Closes all figure windows

%Parameters
a =
b =
Ibar =
Gbar =
```

```
%Intial Condition
Y0 = ;

%Fixed point
Yfp = (a+ Ibar + Gbar)/(1-b);

%Simulation parameters
n= ; % Number of iterations

%Simulation
Y(1:n+1)=zeros(1,n+1);
Y(1) = Y0;
for m = 1:n
    Y(m+1) = b*p(k) + a + Ibar + Gbar;
end

%Plotting
figure(9),hold on
plot(0:n,Y,'ko','MarkerFaceColor','k')
plot([0,n],[Yfp,Yfp],'k--')
xlabel('t'),ylabel('Y_t')
box on
hold off
```

Since we have the analytical solution of difference equation (4.13), it is of course possible to iterate the solution (4.16) directly in the `%Simulation` part of the above code by replacing it with the following code.

The iterate the solution

```
%Simulation
Y(1:n+1)=zeros(1,n+1);
Y(1) = Y0;
for m = 1:n
    Y(m+1)=(Y0-Yfp)b^t+Yfp;
end
```

Exercise: Simulate the Keynesian multiplier model.

Simulate the Keynesian multiplier model and try different values of the parameters a, b, $\bar{I}$ and $\bar{G}$, initial condition Y_0 and iteration parameter n.

4.4 CONCLUDING REMARKS, EXTENSIONS AND CHALLENGES

In this chapter, we studied the Keynesian income determination model. The model we used is linear and quite basic, with a view to understanding the underlying dynamics of the multiplier process. We saw that the Keynesian model is a demand driven system where expenditures of households, firms and governments determine national income via the multiplier. This is one of the fundamental insights of Keynes [5], where he showed how expenditures drive the system to macroeconomic equilibrium, i.e. the saving-investment equality. Keynes' approach ran counter to the Neoclassical economics' view of the savings driven system, where investment passively adjusts to available savings. Keynes reversed the causation and described how investment determines the required level of saving that brings the macroeconomy to equilibrium. Further, Keynes' analysis showed how the savings driven system can exhibit what he called the *paradox of thrift*, where increase in savings leading to weakening of demand and the consequent fall in the equilibrium level of output.

We can analyse the paradox of thrift argument using our model by comparing two different values of b. A large value of b means that agents' propensity to consume (save) is higher (lower), and a small b means that agents' propensity to consume (save) is lower (higher). In Figure 4.2 we simulate the system for two different values of b, with two different initial values Y_0 in each case. In Figure 4.2(a) with $b = 0.2$, where the system

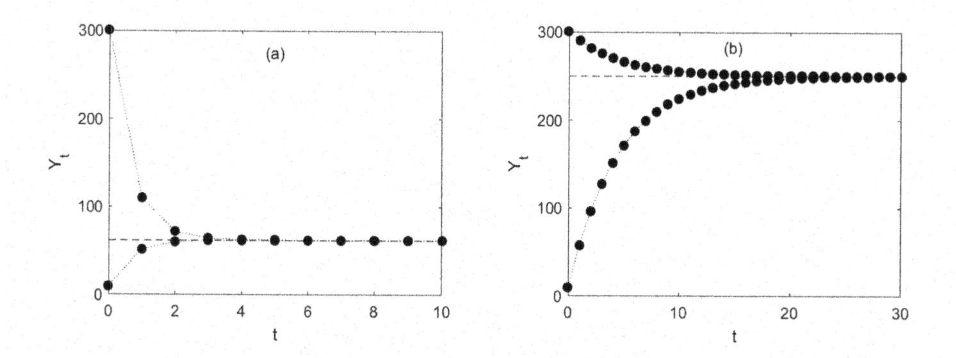

Figure 4.2: Evolution of output Y over time t for the Keyenesian multiplier model (4.13) with parameter values $a = 10$; $\bar{I} = 20$ and $\bar{G} = 20$ and where in (a) $b = 0.2$ and in (b) $b = 0.8$. The equilibrium values are indicated with horizontal dashed lines and the two initial conditions in each case are $Y_0 = 10$ and $Y_0 = 300$.

yields a lower equilibrium value (Eqn. 4.15) than in Figure 4.2(b) where b = 0.8 yields a higher equilibrium value. We also see in Figure 4.2 that when b is low the speed of convergence to the equilibrium value is faster than when b is high where the convergence is slower, which is what we would expect from Eqn. (4.16).

Further, the basic model can also be extended by introducing taxes (denoted by τ) and an independent investment function for firms. These are standard textbook cases, which the readers can explore as simple tasks.

Task 1: Introduce taxes and determine the multiplier.

Verify that the introduction of a lump-sum tax τ in the model and by rewriting the consumption as a function of disposable income yields the multiplier relation:

$$\frac{1}{1-b(1-\tau)}.$$

Task 2: Extending the investment function

Consider a linear investment function given by $I = I_0 - I_1 r$, where $I_0 > 0$ is the autonomous investment and $I_1 > 0$ is the part of investment that relates to the interest rate r, which is the cost of borrowing. Verify that you get the following expression for the equilibrium income:

$$Y^* = \frac{1}{1-b(1-t)}[a_0 + (I_0 - I_1 r) + G] = m[(a_0 + I_0 + G) - I_1 r] = A - ar,$$

where

$$m = \frac{1}{1-b(1-t)}, \quad A = m(a_0 + I_0 + G) \quad \text{and} \quad a = mI_1.$$

Task 3: The goods market equilibrium curve or the *IS* curve

As we will see in the next chapter, the expression $Y^* = A - ar$ in Task 2 is the *goods market equilibrium curve* that is known as the *IS* curve, i.e. the investment-saving equality curve. Readers are encouraged to experiment with different parameter values of I_1 and then study the implications for the equilibrium level of income.

We can also introduce another extension to the investment function by incorporating the "acceleration effect" through the general function

$$I = I(r, Y) \quad \text{with} \quad \frac{dI}{dr} < 0, \quad \frac{dI}{dY} > 0.$$

In this formulation, the derivative $\frac{dI}{dr}$ is assumed to be negative reflecting the adverse effect of the cost of borrowing on investment, but the derivative $\frac{dI}{dY}$ is assumed to be positive reflecting the positive impact increasing output has on investment, which is the so-called "acceleration" effect. Such a formulation introduces interaction between the multiplier effect, on the one hand, and the acceleration effect, on the other, and generates richer dynamics in the Keynesian income determination model. We will study the dynamics of the multiplier-accelerator interaction in detail in Chapter 11.

NOTE

1 See Edmond Malinvaud [3] for a modern introduction to the general equilibrium theory and Alan Kirman's succinct critique [6] of the general equilibrium theory.

BIBLIOGRAPHY

[1] Gordon, R. (1990), What Is New-Keynesian Economics?, *Journal of Economic Literature,* Vol. 28, No. 3, pp. 1115–1171.

[2] Harrod, R. (1937), An Essay in Dynamic Theory, *Economic Journal,* Vol. 49, pp. 14–33.

[3] Malinvaud, E. (2012), Neoclassical General Equilibrium Theory as a Source of Powerful Concepts, Although Confronted with the Complexity of Market Economies, *Œconomia – History, Methodology, Philosophy,* Vol. 2, No. 1, pp. 3–14.

[4] Kalecki, M. (1971), *Selected Essays on the Dynamics of the Capitalist Economy,* Cambridge, UK: Cambridge University Press.

[5] Keynes, J.M. (1936), *The General Theory of Employment, Interest and Money,* London: Macmillan.

[6] Kirman, A. (1986), The Intrinsic Limits of Modern Economic Theory: the Emperor has no Clothes, *The Economic Journal,* Vol. 99, No. 395, pp. 126–139.

[7] Pasinetti, L.L. (1962), Rate of Profit and Income Distribution in Relation to the Rate of Economic Growth, *Review of Economic Studies,* Vol. 29, pp. 267–279.

[8] Robinson, J. (1956), *The Accumulation of Capital,* London: Macmillan.

5 THE IS/LM MODEL

5.1 ECONOMIC PROBLEM

In the previous chapter we considered the Keynesian system without its most important aspect – the role of money. The most fundamental contribution of Keynes' seminal book *General Theory of Employment, Interest and Money* (*General theory*) [7] is the articulation of the commodity production in the capitalist market economies as a monetary theory of production – emphasizing the central role of money in the commodity production. For Keynes, as opposed to the neoclassical economics, money is not just a unit of account and a medium of exchange, but also a store of value. The store of value characteristic of money is what determines interest rates, through the money-demand function, which together with the so-called 'animal spirits' determines investment, output and employment. Whereas in the neoclassical economics, the level of output in the economy is determined by the supply of labour and the demand for labour by the optimizing households and firms, respectively, and is then allocated between consumption and saving by the interest rate. As we discussed in the previous chapter, the available saving (or the supply of loanable funds) leads to investment (or the demand for the loanable funds) and in this sense these models are referred to as the 'supply-side' models. From the macroeconomic saving-investment equilibrium point of view, the neoclassical economists believe that the direction of causation runs from saving to investment. However, Keynes' fundamental contribution in the *General theory* is to show how the direction of causation works in reverse, i.e. investment determining saving in the context of a modern monetary economy.

John R. Hicks, the renowned British economist, compared the two contrasting models in his article titled *Mr. Keynes and the Classics: A suggested interpretation* [5], where he created an analytical apparatus for the purpose, which became known as the *IS-LM model* (Investment and Saving – Liquidity preference and Money supply). Since then the IS-LM model became the main theoretical framework for the exposition of the Keynesian ideas in the class room as well as a popular tool for policy analysis.

DOI: 10.4324/9780429324406-5

Although the IS-LM apparatus appears to be 'Keynesian' in spirit, it has many inconsistencies in terms of representing the analytical core of Keynes' *General theory*. One of the major issues with the IS-LM model is its neglect of the relation between stocks and flows. In the *IS* side of the model, the relation between capacity utilisation of investments undertaken by firms is not considered, while on the *LM* side, the effect of private and public savings over the stock of government bonds is neglected. For instance, Godley and Lavoie [4] argue that the IS-LM model does not specify what the households do with their savings and consequently the flows of income (as savings) bear no relation to the level of existing stocks of financial assets, such as government bonds.

This inconsistency has serious implications for the notion of equilibrium of the IS-LM model. Tobin [10] highlighted how the inconsistency between the relation between stocks and flows leads us to question the stationarity, or the stability of the equilibrium of the model. Tobin argued that the solution of the model in terms of the flow variables implies that the underlying stocks are either increasing or decreasing. For example, a rise in equilibrium level of households' savings implies that their net worth (stock) has increased; similarly, an increase in government deficit (or surplus) implies an increase (decrease) in the stock of public debt. Such stock changes matter because they determine flows, for instance, wealth in the saving function or capital in the production function. Therefore, given this internal dynamics in the IS-LM model, the only way it can display a stable equilibrium is when all the underlying stocks of the flow variables grow at the same proportional rate.

The theoretical problem posed by this inconsistency means that there is no correspondence between the comparative statics of the IS-LM model and the fully fledged dynamic analysis where the relations between flows and stocks are fully specified. For these reasons, it is argued that the IS-LM apparatus does not represent Keynes' original ideas espoused in the *General theory* [7].

The stock-flow inconsistency

Hicks understood the stock-flow inconsistency in the IS-LM model and in a conversation with Arjo Klamer in 1988 he conceded: "I did a sort of revision of IS-LM, not many years ago, but I now feel that I have gotten to the point. It is quite simple. Those two curves do not belong together. One is a flow equilibrium, the other a stock. They have no business being on the same diagram." (see p. 175 of [6])

Notwithstanding these inconsistencies, the IS-LM model remains the popular classroom model for teaching "Keynesian" economics. There is a rich discussion on the history and persistence of the IS-LM model in the undergraduate teaching (see for instance [2]). Since the purpose of this book is limited to understanding the dynamics of some well-known models, we will stick to that objective and look at the dynamics of the IS-LM model in this chapter.

5.2 MODELLING

In setting up the dynamic version of the IS-LM model, we assume that the readers have already been introduced to the basic model. For a preliminary understanding of the IS-LM model, readers are directed to any standard introductory/intermediate level macroeconomics textbooks, such as Mankiw [8] or Dornbusch and Fisher [3]. Let us first introduce the IS and LM curves, and then set up the IS-LM model.

The IS curve. In the following we consider a closed economy with a central bank, but without government. The IS curve represents the goods market equilibrium, where planned expenditure is equal to actual expenditure, and is given by

$$Y = C + I, \tag{5.1}$$

where C is the consumption expenditure and I is the investment expenditure, and yields the equilibrium condition

$$Y - C = I \quad \Rightarrow \quad S = I, \tag{5.2}$$

where S is the total savings in the economy.

Further assuming that both saving S and investment I are determined by the income Y and interest rate r, the goods market equilibrium is expressed as

$$S(Y,r) = I(Y,r), \tag{5.3}$$

where both $S(Y,r)$ and $I(Y,r)$ are nonlinear functions, in general. The IS curve is given by the solution of (5.3), which lives in the (Y,r) plane, and it shows the combination of real income and interest rate that leads to the equilibrium in the goods market.

The LM curve. In terms of the money market, there are two financial assets in the economy, money supplied by the central bank and bonds issued by private firms. The money supplied by the central bank is denoted by M and is assumed here to be constant, and the demand for money in the economy is assumed to be given by $L = L(Y,r)$, which depends on income Y and interest rate r. The transaction and precautionary motives for the money demand is being driven by income (Y) and the speculative motive for the money demand is reflected by the interest rate (r). The transaction and precautionary motives imply that the money demand increases with a rise in income, i.e. money demand L increases with income Y. With regards to the speculative motive, the money demand is inversely related to interest rate, i.e. if bonds offer a higher (lower) interest rate then the demand for money falls (rises).

Let M denote the nominal amount of money supply and P be the general price level, then the money market equilibrium is characterised by

$$\bar{M} = L(Y,r), \tag{5.4}$$

where $\bar{M} = \frac{M}{P}$ represents the supply of real money balances in the economy. The LM curve is given by the solution of (5.4), which lives in the (Y,r) plane and shows the combination of real income and interest rate that leads to the equilibrium in the money market.

The IS-LM model. The macroeconomic equilibrium is obtained when both the goods and money market simultaneously clear, and can formally be written

$$I(Y,r) - S(Y,r) = 0, \quad (\text{IS}) \tag{5.5}$$

$$L(Y,r) - \bar{M} = 0, \quad (\text{LM}) \tag{5.6}$$

where the equations describe two equilibrium curves in the (Y,r) plane. The two curves intersect in a point (Y^*,r^*) – the macroeconomic equilibrium – where the real and the monetary sectors are in balance simultaneously, as shown in Figure 5.1.

In order to study the disequilibrium dynamics, i.e. to find out what happens if the system is not in equilibrium, we need to recast the IS-LM model in a dynamic form. This is where we need to be guided by economic theory. The economy in Keynes's *General theory* [7] is characterised by under full-employment, where firms have unutilised capacity and workers are involuntarily unemployed due to lack of demand. In such an economy, output adjusts through, the so-called *quantity adjustment* for any

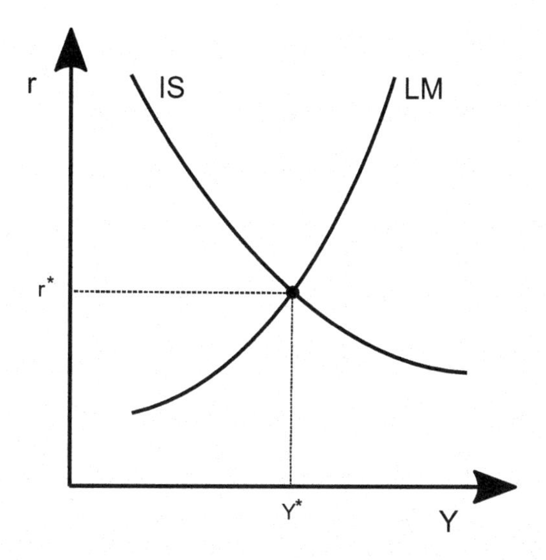

Figure 5.1: The IS/LM model with the two curves, IS and LM, and the equilibrium point (Y^*, r^*) where the curves cross.

excess demand in the product market, i.e. when investment exceeds saving. However, in the money market, where the money supply is fixed by the central bank, excess money demand leads to a *price adjustment,* i.e. an increase in the rate of interest r. Therefore, excess demand triggers quantity adjustment in the product market and price adjustment in the money market. We can formalise this economic intuition using a system of first-order ordinary differential equations as

$$\frac{dY}{dt} = c_1(I(Y,r) - S(Y,r)), \tag{5.7}$$

$$\frac{dr}{dt} = c_2(L(Y,r) - \bar{M}). \tag{5.8}$$

where $c_1 > 0$ and $c_2 > 0$ and referred to as the *speed of adjustment* coefficients. The economic interpretation of the speed of adjustment coefficients is simply the speed with which a particular market adjusts for excess demand (or supply). For instance, the product market may adjust relatively slowly due to the gestation lags of new investment, whereas the money market reacts relatively quickly for any excess demand (or supply) conditions. The reader will be able to experiment with the relative speeds of adjustment later in the simulation exercises and study their effect on the dynamics of the system.

The signs and magnitudes of partial derivatives in the investment I, saving S and money demand L functions with respect to the variables Y and r are assumed as

$$0<\frac{\partial I}{\partial Y}<1, \quad \frac{\partial I}{\partial r}<0, \tag{5.9}$$

$$0<\frac{\partial S}{\partial Y}<1, \quad \frac{\partial S}{\partial r}>0, \tag{5.10}$$

$$\frac{\partial L}{\partial Y}>0, \quad \frac{\partial L}{\partial r}<0, \tag{5.11}$$

where (5.9) says that the investment I increases with income Y and decreases with the interest rate r, the conditions in (5.10) stipulate that savings S increases with both income and the interest rate, and (5.11) states that the money function L increases with income and decreases with the interest rate. Further, note that the magnitude of the "acceleration" effect in (5.9), i.e. the response of investment for any change in income, is less than 1, which is necessary, as we shall see below, for the stability of the equilibrium. Otherwise, the acceleration effect will introduce a self-reinforcing dynamics between investment and output, and together with the multiplier effect it will make the system expand indefinitely.[1] The magnitude of the partial derivative of saving with respect to output (5.10) also provides a necessary condition for the stability of the equilibrium such that it prevents the system from being trapped in excess savings, also referred to as *savings glut*.[2]

Now, using (5.2), i.e. $S(Y,r) = Y - C(Y,r)$, we can rewrite (5.7)-(5.8) as

$$\frac{dY}{dt} = c_1(I(Y,r)+C(Y,r)-Y), \tag{5.12}$$

$$\frac{dr}{dt} = c_2(L(Y,r)-\bar{M}), \tag{5.13}$$

where (5.12) represents the goods market, and thus the goods market equilibrium curve found when $I = S$ $(= C - Y)$, which yields the goods market *nullcline* where $\frac{dY}{dt}=0$. Similarly, from (5.13) we can find the money market equilibrium curve when $L(Y,r) = M$, which is the money market *nullcline* where $\frac{dr}{dt}=0$.

In the following analysis, we will work with specific functional forms for consumption, investment and money demand functions and study the dynamics.

Dimensional consistency

Before we proceed to analyse the system, we would like to alert the readers to be cognizant of the units, or *dimensions,* in which the variables and parameters are defined in the model. A simple definition of *dimension* is that it is a set of all quantities that can be compared, added, or subtracted, etc. For instance, if apples are defined in *money* terms (M) and oranges are defined in terms of *quantities* (Q), then we cannot add the two. However, we can add apples and oranges in terms of their *value*, which is expressed in money units. Similarly, the IS-LM model has variables with different dimensions and therefore one has to verify whether the dynamical system is *dimensionally consistent.* Consider the LM equation (5.13), the right-hand side is the difference between the money demand and money supply variables, which are given in money terms with dimension M. However, the left-hand side is the rate of change in interest rate, whose dimension is $\frac{1}{T^2}$, where T denotes time and the interest rate has dimension $\frac{1}{T}$. Therefore, the left-hand side and right-hand side of equation (5.13) are potentially dimensionally inconsistent. In order to make them consistent, the dimension of the speed of adjustment parameter (c_2) has to be $\frac{1}{MT^2}$ ('per unit of money and (unit of time)^2') [9]. Similarly, the reader can verify the dimension of the speed of adjustment (c_1) of the IS curve (5.12) that renders its dimensional consistency. The issue of dimensional consistency is very important, like the stock-flow consistency, but usually ignored in the undergraduate textbooks in economics.

5.3 ANALYSIS, SIMULATION AND VISUALISATION

The standard representation of the IS-LM model in undergraduate textbooks is to assume linear forms for the consumption function C, the investment function I and the money demand function $L(Y,r)$ as

$$C(Y) = a_1 Y, \qquad 0 < a_1 < 1, \tag{5.14}$$

$$I(r) = -a_2 r + \overline{I}, \quad a_2 > 0, \tag{5.15}$$

where $\overline{I}$ is *autonomous investment*, a_1 and a_2 are two positive constants, and thus substituting (5.14) and (5.15) in (5.12) we get

$$\frac{dY}{dt} = c_1(a_1 Y - a_2 r + \overline{I} - Y) = -c_1 s Y - c_1 a_2 r + c_1 \overline{I}, \quad 0 < s < 1, \tag{5.16}$$

where $s = 1 - a_1$.

Next, let us turn to the money demand function $L(Y,r)$ and assume that the money demand is a linear increasing function of income Y and decreasing with respect to interest rate r, so that

$$L(Y,r) = k_1 Y - k_2 r, \quad k_1 > 0, k_2 > 0, \tag{5.17}$$

where k_1 and k_2 are two positive constants.

The money supply $\bar{M}$ is determined by the Central bank and assumed to be constant. Substituting (5.17) and the assumption on $\bar{M}$ in equation (5.13), which defines the rate of change in interest rate, yields

$$\frac{dr}{dt} = c_2 k_1 Y - c_2 k_2 r - c_2 \bar{M}. \tag{5.18}$$

Equations (5.16) and (5.18) represent the IS-LM model in the dynamic form with specific functions of consumption, investment and money demand, and describe the dynamics of output and interest rate in a simple closed economy. Since both (5.16) and (5.18) are linear we can analyse the system analytically and then perform experimentation of parameter variations using the numerical simulation.

First, for the analysis, let us rewrite equations (5.16) and (5.18) in matrix form,[3] so that

$$\begin{pmatrix} \frac{dY}{dt} \\ \frac{dr}{dt} \end{pmatrix} = \begin{pmatrix} -c_1 s & -c_1 a_2 \\ c_2 k_1 & -c_2 k_2 \end{pmatrix} \begin{pmatrix} Y \\ r \end{pmatrix} + \begin{pmatrix} c_1 \bar{I} \\ -c_2 \bar{M} \end{pmatrix}. \tag{5.19}$$

The equilibrium point for this system, i.e. where $\frac{dY}{dt} = 0$ and $\frac{dr}{dt} = 0$ simultaneously, is given by

$$\begin{pmatrix} \frac{dY}{dt} \\ \frac{dr}{dt} \end{pmatrix} = \begin{pmatrix} 0 \\ 0 \end{pmatrix} \Rightarrow \begin{pmatrix} -c_1 s & -c_1 a_2 \\ c_2 k_1 & -c_2 k_2 \end{pmatrix} \begin{pmatrix} Y^* \\ r^* \end{pmatrix} + \begin{pmatrix} c_1 \bar{I} \\ -c_2 \bar{M} \end{pmatrix} = \begin{pmatrix} 0 \\ 0 \end{pmatrix}$$

and thus the equilibrium level of income Y^* and interest rate r^* yields

$$\begin{pmatrix} Y^* \\ r^* \end{pmatrix} = \frac{1}{s k_2 + a_2 k_1} \begin{pmatrix} k_2 \bar{I} + a_2 \bar{M} \\ k_1 \bar{I} - s \bar{M} \end{pmatrix} = \frac{1}{s k_2 + a_2 k_1} \begin{pmatrix} k_2 & a_2 \\ k_1 & -s \end{pmatrix} \begin{pmatrix} \bar{I} \\ \bar{M} \end{pmatrix}.$$

The stability of the equilibrium point (Y^*, r^*) is determined by the eigenvalues λ_1 and λ_2 of the *coefficient matrix*

$$A = \begin{pmatrix} -c_1 s & -c_1 a_2 \\ c_2 k_1 & -c_2 k_2 \end{pmatrix}, \tag{5.20}$$

which is given in (5.19). The two eigenvalues are found directly by calculating the roots of the corresponding characteristic equation

$$\begin{vmatrix} -c_1 s - \lambda & -c_1 a_2 \\ c_2 k_1 & -c_2 k_2 - \lambda \end{vmatrix} = \lambda^2 + (c_1 s + c_2 k_2)\lambda + c_1 c_2 (s k_2 + a_2 k_1) = 0 \quad (5.21)$$

and thus they are

$$\lambda_{1,2} = \frac{1}{2}\left(-(c_1 s + c_2 k_2) \pm \sqrt{(c_1 s - c_2 k_2)^2 - 4 c_1 c_2 a_2 k_1} \right).$$

Note, the characteristic equation is nothing but $\lambda^2 - \tau\lambda + \Delta = 0$, where τ and Δ are respectively the trace and determinant of the coefficient matrix A. Since c_1, c_2, s, a_2, k_1, k_2 are all positive, we can directly infer from (5.20) that $T < 0$ and $\Delta > 0$, and thus

$$\begin{aligned} -\tau &= c_1 s + c_2 k_2 > 0, \\ \Delta &= c_1 c_2 (s k_2 + a_2 k_1) > 0, \end{aligned}$$

which are the sufficient conditions for stability of the equilibrium point.

Having found the equilibrium point and determined its stability conditions analytically, we will next introduce MATLAB® code for numerical simulation of the IS-LM model and study and visualise its dynamics. To do this we need to specify the parameter values and initial conditions for the analysis of the system.

Parameter values and initial conditions

To see how the IS-LM model behaves we use the following numerical values for the parameters

$$c_1 = 1,\ c_2 = 1,\ s = 0.5,\ a_2 = 1,\ k_1 = 1,\ k_2 = 0.1,\ \bar{I} = 0.55,\ \bar{M} = 1$$

and initial conditions

$$Y_0 = 1,\ r_0 = 0.01.$$

We note that this set of values are only used as an example. The readers are encouraged to test different parameter values (even values that do not make economic sense) in order to get a better feeling for dynamics of the system.

The MATLAB® code to simulate the system of two first-order differential equations (5.19) is given below in five separate parts, namely, (i) the initiation of the simulation and parameter values, (ii) the initial conditions, the call of the differential equation solver ode45[4] and the output, (iii) the IS-LM system, (iv) the calculation of the equilibrium point and the eigenvalues, and (v) the plotting of the time histories of $Y(t)$ and $r(t)$, and the trajectory $(Y(t), r(t))$.

The first set of code clears the memory, closes all the figures and sets the parameters.

Initiation and parameter values

```
%IS-LM model
clear all
close all
format compact

%Parameters
c1 = 1;
c2 = 1;
s = 0.5;
a2 = 1;
k1 = 1;
k2 = 0.1;
Ibar = 0.55;
Mbar = 1;
A = [-c1*s, -c1*a2; c2*k1, -c2*k2]; % The coefficient matrix
b = [c1*Ibar; -c2*Mbar]; % The constant vector
```

Next we set up the simulation that starts at time $t = 0$ and runs for 200 time units, i.e. $t = 200$, with initial conditions $(Y_0, r_0) = (1, 0.01)$. In MATLAB® one can chose between different numerical solvers to solve ordinary differential equations (ODEs), and here we have picked a solver named ode45. Through the options structure, we can set solver properties such as error tolerances, i.e. the maximum size of the errors of the numerical ODE solution. The output for the ODE solver is a time vector t and 2-column state matrix Z, where the first column is Y and the second column is r. Finally, we separate the output in a Y and r vector to be used in the plotting.

Initial condition and the ODE solver

```
%Simulation time
t_start = 0;
t_end = 200;

%Initial conditions
Y0 = 1;
r0 = 0.01;

%Simulation tolerances
options = odeset('RelTol',1e-6,'AbsTol',1e-6);
[t,Z] = ode45(@(t,y) islm_system(t,y,A,b),...
              [t_start t_end],[Y0;r0],options);

%Output
Y = Z(:,1);
r = Z(:,2);
```

The ODE solver `ode45` calls the function `islm_system(t,y,A,b)`, which evaluates the right-hand side of (5.19), with time `t`, state variables `y` and parameters matrix `A` and vector `b`, as input. The code for the `islm_system(t,y,A,b)` function shown below is typically placed at the end of the MATLAB® file, after the plotting instructions and other calculations (see Appendix B). Note that A and b are defined outside of the function `islm_system(t,y,A,b)` and can therefore be updated without having to change the `islm_system(t,y,A,b)` function module.

The IS-LM system

```
function dYdr = islm_system(t,y,A,b)

%Variables
Y = y(1);
r = y(2);

%The ODE system
dYdr = A*[Y;r]+b;

end
```

In order to locate the equilibrium point (Y^*,r^*) and calculate the eigenvalues of the coefficient matrix (5.20) we do not need to simulate the system, but instead we can use the following MATLAB® code.

Equilibrium and eigenvalues

```
%Equilibrium point
Yr_eq = A\(-b); %same as inv(A)*b
Y_eq = Yr_eq(1);
r_eq = Yr_eq(2);

%Eigenvalues
eigenvalues = eig(A);
lambda_1 = eigenvalues(1)
lambda_2 = eigenvalues(2)
```

To visualise the output of the simulation, we can use a similar code that was used in Chapters 2–4. However, note the use of subplot for plotting the $Y(t)$ and $r(t)$ curves separately in Figure 5.2(a) and (b), and the IS and LM curves in Figure 5.2(c).

Plotting the output

```
%Plotting
figure(1)
subplot(2,1,1)
plot(t,Y,'k','LineWidth',2)
axis([0,20,0.98,1.04])
ylabel('Y(t)')
subplot(2,1,2)
plot(t,r,'k','LineWidth',2)
axis([0,20,0,0.08])
xlabel('time (t)'),ylabel('r(t)')

Y_LM = [1,1.008];
r_LM = k1/k2*Y_LM - Mbar/k2;
Y_IS = [0.98,1.04];
r_IS = Ibar/a2-s/a2*Y_IS;
figure(2)
hold on
plot(Y,r,'k','LineWidth',2)
```

```
plot(Y_eq,r_eq,'ko','MarkerSize',6,...
     'MarkerEdgeColor','k','MarkerfaceColor','k')
plot(Y0,r0,'ko','MarkerSize',6,...
     'MarkerEdgeColor','k','MarkerfaceColor','k')
plot(Y_IS,r_IS,'k','MarkerSize',4)
plot(Y_LM,r_LM,'k','MarkerSize',4)
axis([0.98,1.04,0,0.08])
xlabel('Y'),ylabel('r')
hold off, box on
```

After running the entire file, see Appendix B.5, we first get the following output in the MATLAB® workspace:

```
lambda_1 =
  -0.3000 + 0.9798i
lambda_2 =
  -0.3000 - 0.9798i
```

We see that the system has two complex conjugate eigenvalues

$$\lambda_1 = -0.3000 - 0.9798\mathbf{i}, \qquad \lambda_2 = -0.3000 + 0.9798\mathbf{i}$$

and we know from Appendix A.5 that if the real parts of the complex conjugate eigenvalues are negative, then the equilibrium point is stable. Furthermore, since the eigenvalues are complex conjugate we can expect oscillations or spiraling motion around the equilibrium point, while at the same time getting closer to it as time passes. After running the entire MATLAB® code, we also get three figures as output. In Figures 5.1(a) and (b) we see time evolutions of the income $Y(t)$ and interest rate $r(t)$ and notice the oscillations, which are as expected due to the complex conjugate eigenvalues. In Figure 5.2(c) the interest rate r is plotted against the income Y, where we see the counter-clockwise spiralling motion corresponding to the oscillations in Figure 5.2(a) and (b). In Figure 5.2(c) we have also plotted the IS and LM curves given by

$$\begin{aligned} sY + a_2 r - \bar{I} &= 0, \quad \text{(IS)} \\ k_1 Y - k_2 r - \bar{M} &= 0, \quad \text{(LM)} \end{aligned}$$

which are derived from (5.5), (5.6), (5.16) and (5.18).

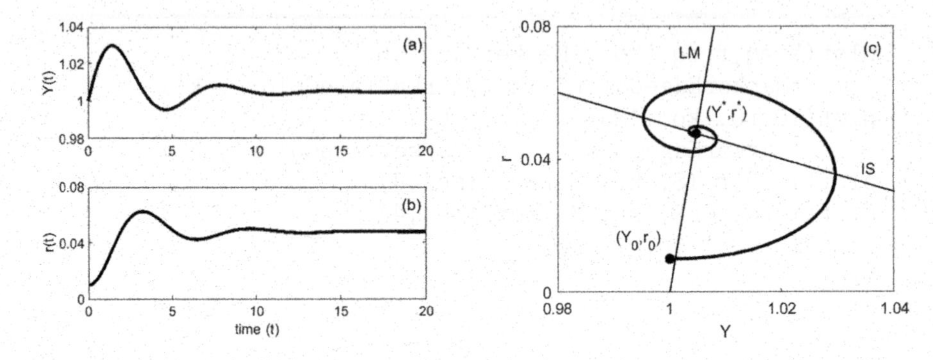

Figure 5.2: Evolution of (a) the income Y and (b) the interest rate r versus time t of the system described in (5.19), and in (c) the income Y is plotted versus interest rate r. In all three plots the parameter values used are $c_1 = 1$, $c_2 = 1$, $s = 0.5$, $a_2 = 1$, $k = 1$, $\bar{I} = 0.55$ and $\bar{M} = 1$. The curves $Y(t)$ and $r(t)$, the initial conditions (Y_0,r_0) and the equilibrium point (Y^*,r^*) are indicated in (c).

5.4 REMARKS, EXTENSIONS AND CHALLENGES

In this chapter we set up a dynamic IS-LM model (5.19), which is ready to be used for further experimentation and analysis. At the minimum, one can think of three different scenarios for experimentation. First, we can conduct policy experimentation, in terms of both fiscal and monetary policy. For instance, to introduce an expansionary monetary policy, we increase the money supply, i.e. $\bar{M}$ in (5.18), which will shift the LM curve to the left, and study the resulting disequilibrium dynamics of the system. As we know from the stability conditions, the system will "spiral" towards a new equilibrium, which is a lower-output and higher-interest-rate equilibrium. The spiralling dynamics means that both output and interest rate fluctuate in the short run as they converge to the new equilibrium. The policy question is how one could reduce the fluctuations, which are costly since they introduce some uncertainty in the minds of investors, and make the system converge relatively monotonically towards a new equilibrium. This is where the role of the 'speed of adjustment' parameters (c_1,c_2) comes into play, which leads to the second experimentation task.

Task 1: Policy analysis

Experiment with various policy options using the system (5.19) by (i) varying the money supply ($\bar{M}$) and the autonomous investment spending ($\bar{I}$) values; (ii) changing the slope of the IS and LM curves (the parameters s, a_2, k_1 and k_2) and study the effectiveness of an expansionary fiscal (or monetary) policy; (iii) simulating scenarios that maintain the interest rate at a particular level when an expansionary monetary policy takes place, or keeping the economy at a particular level of income when an expansionary fiscal policy is undertaken.

In the above analysis of the numerical example, we have assumed that both the goods market and money market have equal speed of adjustment, by assuming the parameter values as $c_1 = c_2 = 1$. In reality, the goods market adjusts more slowly than the money market. Therefore, one could experiment with the relative speeds of adjustment in the above simulation. One possibility is to make the money market adjustment much faster, relatively to the goods market, and explore the disequilibrium dynamics. In the extreme scenario, the money market clears instantaneously. In such a scenario, for instance when money supply is decreased, the dynamics of the system 'jumps' from the original equilibrium onto the LM curve and then moves along the curve to reach the new equilibrium, as the goods market slowly adjusts to monetary shock. In this case, the speed of adjustment parameter c_2 introduces a *discontinuity* in the disequilibrium dynamics of the system.[5] While this is an extreme scenario, readers are encouraged to experiment with varying the relative speeds of adjustment, i.e. the parameters c_1 and c_2, and study the resulting dynamics.

Task 2: Disequilibrium dynamics

Study the disequilibrium dynamics of the system by experimenting with the relative speeds of adjustment parameters c_1 and c_2. For instance, keep c_1 constant and simulate the system for three different values of c_2 and visualise the spiraling dynamics for the three cases.

There are many ways one could extend the IS-LM model. The basic textbook model used here ignores the role of expectations, which affects investment, money demand and possibly other functions as well. The role of expectation can be explicitly introduced by considering nominal and real interest rates in the money demand and investment functions, respectively. As we saw in Chapter 4, the specification of the expectation rule, adaptive or otherwise, will impact the nature of the spirals or fluctuations in output and interest rate.

NOTES

1 The interaction between the accelerator and multiplier generates interesting dynamics in terms of business cycle fluctuations and we will study these aspects in Chapter 11.
2 Interestingly, the "global savings glut" due to excess saving in the emerging market economies was argued as one of the main reasons for the Great Financial Crisis of 2008 that was witnessed in the developed western economies [1].
3 The general method of solving a system of equations is discussed in Appendix A.5.
4 An explicit (4,5) Runge-Kutta solver [11].
5 See Shone [9] for a qualitative analysis of this issue.

BIBLIOGRAPHY

[1] Bernanke, B. (2007), Global imbalances: recent developments and prospects. *Bundesbank Lecture speech,* September 4, p. 18.
[2] Colander, D. (2003), The Strange Persistence of the IS-LM Model. *History of Political Economy,* 36(1), pp. 305–322.
[3] Dornbusch, R. and Fisher, S. (2018), *Macroeconomics (12th edition)*, McGraw Hill Education.
[4] Godley, W. and Lavoie, M. (2006), *Monetary economics: an integrated approach to credit, money, income, production and wealth*, Springer.
[5] Hicks, J.R. (1937), Mr. Keynes and the 'Classics'; a Suggested Inter-Pretation, *Econometrica,* pp. 147–159.
[6] Klamer, A. (1989), An Accountant among Economists: Conservations with Sir John R. Hicks. *Journal of Economic Perspectives,* Vol. 3, No. 4, pp. 167–180.
[7] Keynes, J.M. (1936), *The General Theory of Employment, Interest and Money*, London: Macmillan.

[8] Mankiw, N.G. (2019), *Macroeconomics (10th edition)*. London: Macmillan.
[9] Shone, R. (2002), *Economic Dynamics: Phase Diagrams and their Economic Application*, Cambridge University Press.
[10] Tobin, J. (1982), *Asset Accumulation and Economic Activity: Reflections on Contemporary Macroeconomic Theory*, University of Chicago Press.
[11] Shampine, L.F. and Reichelt, M.W. (1997), The MATLAB® ODE Suite, *SIAM Journal on Scientific Computing*, Vol. 18, pp. 1–22.

6 DEBT, DEFICIT AND STABILISATION POLICY

In broad terms, fiscal policy is concerned with the role governments play in the market economy. Fiscal policy is a tool for governments to stabilise the economy when markets crash, to stimulate the economy during recessions, and to achieve their wider social objectives such as redistribution of income between different sections of society and the provision of public goods. Governments use a progressive (income) tax system and the transfers such as unemployment benefits and social security benefits to redistribute income across the income groups. This aspect of redistribution is one of the main reasons why fiscal policy remains more contentious than monetary policy. Fiscal policy also plays a vital role in the provision of public goods, which are non-excludable and non-rivalrous such as military, clean water etc.

The role of fiscal policy becomes more crucial particularly in those times when monetary policy becomes ineffective. For example, following the financial crash of 2007–2008, the conventional monetary policy hit the zero lower bound, where the nominal interest rate set by the central bank is almost equal to zero and became ineffective for stimulating the economy. The central banks had to resort to unconventional monetary policies such as *quantitative easing*, where the central banks purchased various financial assets from the commercial banks with a view to stimulating the economy through enhancing the availability of credit to firms and households. But in the absence of demand in the economy, a mere availability of credit need not necessarily guarantee a pick-up of funds by investors and stimulate economic growth, which was the case in many European economies that were experiencing sluggish recovery post 2008 crash. In these circumstances, fiscal policy became more important, or was the only tool available for governments to stimulate the economy. However, given the unique characteristic of the 2007–2008 crisis where governments had to bailout big or systemically important banks, fiscal stimulus was based more on austerity policies. This may sound contradictory and one may wonder how one could stimulate the economy via contractionary austerity policies. The logic was

DOI: 10.4324/9780429324406-6

that reducing the debt overhang for governments would bring more confidence in the minds of the lenders and ease credit conditions for the private sector to borrow in the international debt markets.

The financial crash of 2007–2008 and the subsequent recession, referred to as the *Great recession*, have highlighted the economic consequences of high levels of government debt (particularly in the USA and Europe) and brought intense debate on debt, deficits and fiscal policy. In this chapter, we will look at the dynamics of debt and deficits in a simple model and analyze and discuss various policy scenarios.

6.1 ECONOMIC PROBLEM

Fiscal policy is conducted through discretionary and automatic spending measures. Governments use discretionary spending measures when the economy plunges into recession or crisis. The discretionary spending measures, be it tax cuts or an increase in expenditure, are often unplanned and have a direct impact on the fiscal or budget deficits. On the other hand, the automatic component, known as *automatic stabilisers,* is a mechanism that allows increases in spending (e.g. unemployment benefits, tax credits, food stamps etc.) that are built in the government budgets that come into play without a need for any vote from the legislators during an economic slowdown. In other words, the automatic stabilisers refer to the changes in the budget deficit that occur automatically as a response to an economic slow down or crisis. Depending on the severity of recession and crises, governments use these two components to stabilise and stimulate the economy. In the case of the USA, under the American Recovery and Reinvestment Act (ARRA) of 2009, which was the fiscal stimulus program in response to the Great recession of 2008–2009, the automatic stabilisers provided half of total stabilisation and the other half was administered through the discretionary spending. The relation between these two different fiscal policy concepts can be seen in simple terms as follows:

Discretionary fiscal spending = Budget deficit – Automatic stabilisers

At any point of time, if the economy enters recession, i.e. when the current output is below the equilibrium or potential level of output, the automatic stabilisers kick in. If it is absorbed in the budget deficit, i.e. if the two cancel out, then there is no increase in the discretionary fiscal stimulus in the economy. However when budget deficit is greater than the automatic stabiliser component, the additional spending amounts to the discretionary fiscal stimulus. Since the discretionary fiscal spending is driven by the recessionary state of the economy, the deficit thus caused is known as the *cyclically adjusted budget deficit* or *structural budget deficit*. In the context of the Covid-19 crisis, many governments announced discretionary fiscal stimulus and pledged more fiscal support through fiscal measures to aid economic recovery. Such stimulus will undoubtedly will have a huge impact on their budget deficits, not just in the current period but also in the near future. There are serious concerns about debt sustainability and its implications for future growth prospects. Some economists argued that countries with a debt-to-GDP ratio above 90 per cent are not sustainable and make economies vulnerable to crisis [2]. However, others argue that there is no such critical threshold and as long as the debt can stimulate economic growth it can be sustainable. We develop a simple model to study the debt-deficit dynamics and study the conditions for debt sustainability.

6.2 MODELLING

Consider an economy with a government that is closed to foreign trade.[1] In order to finance its expenditure, the government can use taxation, sell new bonds or print new money. We assume here that government debt is the stock of government bonds that has been sold to the private sector. The government must continuously finance its current expenditure plan and also pay down its outstanding debt. Equating the uses and sources of funds by the government yields the *budget identity of government* and is given by

$$G + iD = T + \dot{D} + \dot{M}, \tag{6.1}$$

where $G = G(t)$ is the *government expenditure* (in nominal terms) on goods and services, $i = i(t)$ is the *nominal interest rate*, $D = D(t)$ denotes the *outstanding stock of debt* (bonds, in this case), $T = T(t)$ denotes *tax revenues net of transfers* and $M = M(t)$ denotes *central bank money*. Also, $\dot{D} = \frac{dD}{dt}$ denotes the rate at which new bonds are issued and $\dot{M} = \frac{dM}{dt}$ denotes the rate at which new central-bank money is issued by the government.

To further simplify the model, we assume that the government does not have the option of printing new money to finance its deficit, which means that $\dot{M} = 0$. Therefore, the government has to finance its expenditures through taxation or by issuing new bonds (taking on more debt) and, thus, the budget identity (6.1) becomes

$$G + iD = T + \dot{D}. \tag{6.2}$$

Rearranging the identity (6.2) to see the relationship between current budget deficit (the new bonds) and the stock of debt as

$$\dot{D} = G - T + iD, \tag{6.3}$$

which shows that the change in debt equals the current budget deficit plus the interest on outstanding debt. The current budget deficit, excluding interest payments on the debt, is called the *primary deficit*. We express (6.3) as a per cent of GDP by dividing with the *nominal GDP py*, where p (= $p(t)$) is the *price level* and y (= $y(t)$) the *real national income*, to yield

$$\frac{\dot{D}}{py} = \frac{G - T}{py} + \frac{iD}{py} = b + id, \tag{6.4}$$

where $b = \frac{G-T}{py}$ is the ratio of *primary deficit to GDP*, and $d = \frac{D}{py}$ is the *debt to GDP ratio*. Using the debt-GDP ratio, the total value of debt at any point in time is given by

$$D = dpy. \tag{6.5}$$

Differentiating (6.5) with respect to time yields

$$\dot{D} = py\dot{d} + dy\dot{p} + dp\dot{y}. \tag{6.6}$$

And dividing (6.6) by the nominal GDP py we get

$$\frac{\dot{D}}{py} = \frac{py\dot{d}}{py} + \frac{dy\dot{p}}{py} + \frac{dp\dot{y}}{py} = \dot{d} + d\frac{\dot{p}}{p} + d\frac{\dot{y}}{y} = \dot{d} + d\pi + dg, \tag{6.7}$$

where $\pi = \frac{\dot{p}}{p}$ is the rate of inflation and $g = \frac{\dot{y}}{y}$ is the rate of growth of output (GDP).

Equating (6.4) and (6.7), and incorporating the *Fisher equation*

$$r = i - \pi, \tag{6.8}$$

which states that the real interest rate r is equal to the nominal interest rate i minus *inflation* π, we find the dynamic relation between debt and deficit as

$$\dot{d} = (i - \pi - g)d + b = (r - g)d + b. \tag{6.9}$$

Equation (6.9) is the dynamical system that describes the relationship between the rate of change in debt to GDP ratio ($\dot{d}$) and its level (d) along the real interest rate r, the growth of real GDP g and the primary deficit to GDP ratio b.

6.3 ANALYSIS, SIMULATION AND VISUALISATION

In the analysis of (6.9) we will assume that the timescale at which the primary deficit b, the real interest rate r and the growth of real GDP g vary is much slower than the debt-GDP ratio d. In practice, this means that we can consider b, r and g as constants and thus treat

$$\dot{d} = (r - g)d + b \tag{6.10}$$

as a linear first-order ordinary differential equation and analyse using the methods introduced in the previous chapters.

First, from (6.10) we find that the equilibrium point d^* is given by

$$\dot{d} = 0 \quad \Rightarrow \quad b + (r - g)d^* = 0 \quad \Rightarrow \quad d^* = -\frac{b}{r - g} = \frac{b}{g - r} \tag{6.11}$$

and note that d^* is not defined if $r - g = 0$ (assuming $b \neq 0$). From (6.10) we also note that the equilibrium point d^* is stable if $r - g < 0$ and unstable if $r - g > 0$ (for further discussion on the stability properties of first-order differential equations, see Appendix A.3). Note that b is the primary balance of the government and, depending on whether it is in surplus or deficit, we can study the dynamics of (6.10) through four different cases depending on whether the equilibrium d^* is either stable or unstable and the government is either in surplus ($b < 0$) or in deficit ($b > 0$), as shown in Figures 6.1–6.4.

In each of the four cases, we show both the phase-plane plots in the (d, $\dot{d}$) plane to visualise stable (unstable) dynamics of the equilibria and the corresponding time histories of converging (diverging) trajectories, starting from an arbitrary initial condition. At any given level of debt-GDP ratio d, the phase plane plots show how it evolves. The arrows in the phase planes point towards or away from the equilibrium value. We first consider the stable debt-dynamics case where the growth rate g is greater than the real interest rate r and analyse two scenarios where the government's primary balance is positive (deficit) and negative (surplus), as shown in Figures 6.1 and 6.2, respectively. We have used numerical values for parameters for illustrative purposes and they do not reflect real-world data (it is of course possible to use the real economic data here to simulate a real scenario).

Case I. In Figure 6.1 we have the case where growth exceeds the real interest rate, and thus $r - g < 0$, with the government's primary balance being in deficit, i.e. $b > 0$. In this case, the debt-deficit model (6.10) has a stable equilibrium, with the equilibrium value $d^* = 1$. The arrows in the phase-plane plot show the direction of the dynamics, in this case towards the equilibrium value, meaning that any perturbation away from the equilibrium point will bring the system towards the equilibrium value. We see the stability of the equilibrium clearly in the time histories plot in Figure 6.1(b) where we consider two different initial values of the debt-deficit ratios ($d = 1.8$ and $d = -0.8$) and show the evolution of the debt-GDP ratio towards the equilibrium value of $d^* = 1$. In economic terms, this means that as long as the economic growth rate exceeds the real interest rate, the growth of debt is dampened by the fact that the higher growth in the economy more than offsets any increase in interest costs. In other words, the growth in debt is sustainable as long as the economy grows more than the interest cost on its debt.

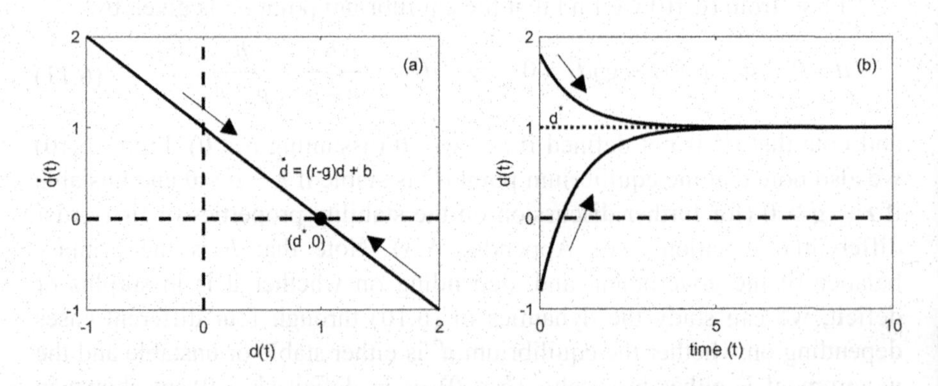

Figure 6.1: The case where $b = 1$, $r = 2$ and $g = 3$ that yields $d^* = 1 > 0$. In (a) d versus $\dot{d}$ is plotted and in (b) time histories for two trajectories with initial condition $d(0) = -0.8$ and $d(0) = 1.8$ are shown, which highlight that $d^* = 1$ is stable. The solid curves correspond to the dynamics of $d(t)$, the dashed lines correspond to $d = 0$ and $\dot{d} = 0$ and the dotted line corresponds to the equilibrium value.

Case II. The argument in Case I holds even in the case when the government's primary balance is in surplus, i.e. when $b < 0$. This case is shown in Figure 6.2. Like in the previous case the debt-deficit equilibrium is stable (since $r - g < 0$), and the only difference in this case is that the economy will converge towards a negative debt-GDP ratio ($d^* = -1$), which means that the government is the net holder of private-sector financial assets.

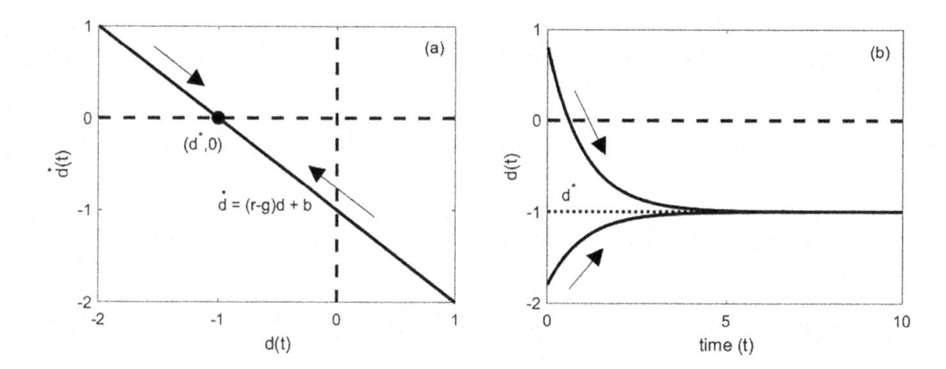

Figure 6.2: The case where $b = -1$, $r = 2$ and $g = 3$ that gives $d^* = -1 < 0$. In (a) d versus $\dot{d}$ is plotted and in (b) time histories for two trajectories with initial condition $d(0) = -1.8$ and $d(0) = 0.8$ are shown, which highlight that $d^* = -1$ is stable. The solid curves correspond to the dynamics of $d(t)$, the dashed lines correspond to $d = 0$ and $\dot{d} = 0$ and the dotted line corresponds to the equilibrium value.

Case III. If instead the interest rate is greater than the growth rate of the economy ($r - g > 0$), the equilibrium point (6.10) becomes unstable. The economic intuition behind the dynamics about the unstable equilibrium is straightforward since the cost of interest on debt increases more than the rate of growth, and the debt-GDP ratio will grow monotonically, as shown in Figure 6.3. Note that the government's primary balance is positive ($b > 0$), i.e. it is in deficit. For the particular parameter values considered here, the equilibrium value of debt-GDP ratio is the same as in Case II, i.e. $d^* = -1$, as shown in Figure 6.3, but now it is unstable. As can be seen from Figure 6.3(b), starting from the two different initial values ($d = -0.8$ and $d = -1.2$) the debt-GDP ratio diverges away from its equilibrium value. The intuition for the monotonic growth of debt in this case is clearly driven by the positive primary deficit of government ($b > 0$). Note that even in the case where the negative debt-GDP ratio, for the initial value of $d = -0.8$, which implies that the economy is a net creditor, the system shows continuous growth of debt after an initial period of surplus. The reason being that the cost of borrowing to maintain or reduce the primary deficit more than offsets the benefits of the initial value of negative debt-GDP ratio (think of it as interest owed on debt vis-a-vis interest earned on credit) and makes the debt-GDP grow and diverge away from the equilibrium. On the contrary, when the initial value of the negative debt-GDP ratio is well below the equilibrium value (in Figure 6.3(b) the initial value $d = -1.2$), the debt-GDP ratio falls monotonically since the benefits of the initial negative debt (or credit) more than outweigh the primary deficit of the government.

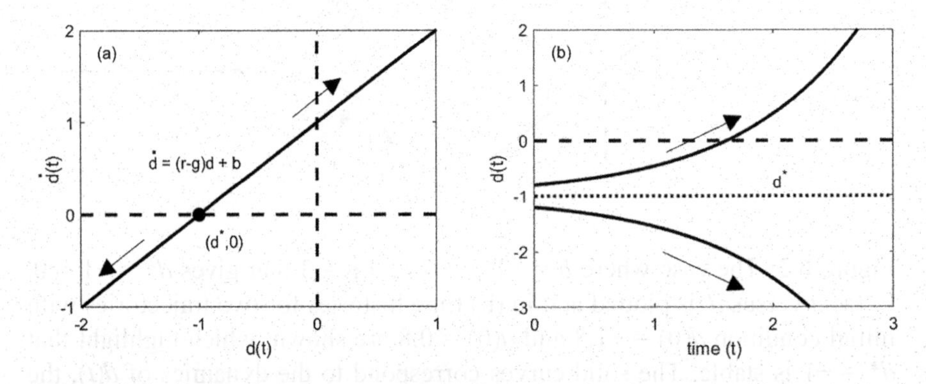

Figure 6.3: The case where $b = 1$, $r = 3$ and $g = 2$ that gives $d^* = -1 < 0$. In (a) d versus $\dot{d}$ is plotted and in (b) time histories for two trajectories with initial condition $d(0) = -0.8$ and $d(0) = -1.2$ are shown, which highlight that $d^* = -1$ is unstable. The solid curves correspond to the dynamics of $d(t)$, the dashed lines correspond to $d = 0$ and $\dot{d}$ and the dotted line corresponds to the equilibrium value.

Case IV. The instability of the system cannot be salvaged by simply making the government's primary balance negative ($b < 0$), i.e. when the government's balance is in surplus. This case is shown in Figure 6.4 and the equilibrium value of the debt-GDP ratio is positive ($d^* = 1$) and unstable, meaning that the system (6.10) will never converge towards the equilibrium. The debt-GDP ratio can grow without bounds for some initial conditions and fall without bounds for others. As discussed in Case III, depending on the size of the primary surplus $b < 0$ relative to the initial debt-GDP value, the system can yield a continuous fall in the debt-GDP ratio as shown in Figure 6.4(b). For instance, when the initial value of the debt-GDP ratio $d = 1.2$, the dynamics of the system yields a monotonic growth in debt. However, when the initial debt-GDP ratio $d = 0.8$, which is less than the primary surplus ($b = -1$), the time history of the trajectory shows a monotonic fall in the debt-GDP ratio. Note that we are analysing a linear system and do not study the feedback between the government's primary balance on growth. Therefore, the mathematical possibility of having such high government primary surpluses may not be economically feasible or viable in reality.

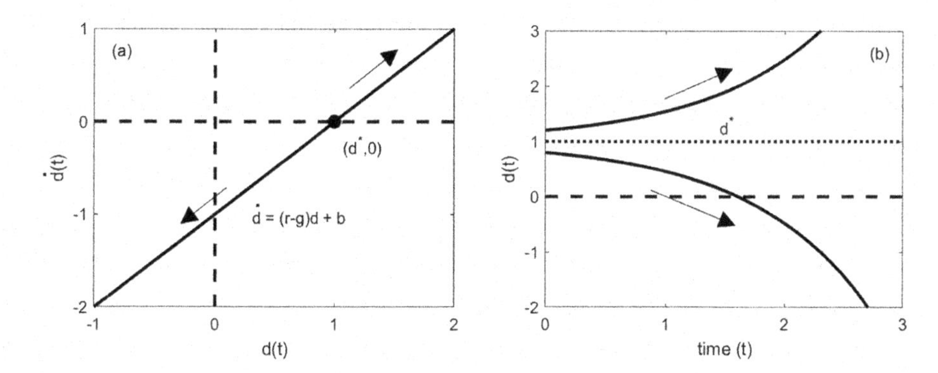

Figure 6.4: The case where $b = -1$, $r = 3$ and $g = 2$ that gives $d^* = 1 > 0$. In (a) d versus $\dot{d}$ is plotted and in (b) time histories for two trajectories with initial condition $d(0) = 0.8$ and $d(0) = 1.2$ are shown, which highlight that $d^* = 1$ is unstable. The solid curves correspond to the dynamics of $d(t)$, the dashed lines correspond to $d = 0$ and $\dot{d} = 0$ and the dotted line corresponds to the equilibrium value.

6.3.1 SIMULATION AND VISUALISATION

Now that we have analytically understood the debt-deficit model (6.10) in terms of its equilibrium and stability properties, we can now simulate the model and experiment with various parametric variations. To simulate the linear first-order ordinary differential equation (6.10) we can use a similar methodology as in Chapter 5. First we set the scene with the following MATLAB® code, where the parameters correspond to Case I:

Initiation and parameter values

```
%debt-deficit
clear all
close all
format compact
```

```
%Parameters
b = 1;
r = 2;
g = 3;

p = [b,r,g]; %Vector of parameters
```

After this initialisation we set up the model for simulation.

Simulation

```
%Simulation time
t_start = 0;
t_end = 10;

%Initial conditions
d0 = 1;

%Simulation tolerances
options = odeset('RelTol',1e-6,'AbsTol',1e-6);
%Simulation tolerances
[t,d] = ode45(@(t,y) debt_deficit_system(t,y,p),...
          [t_start t_end],d0,options);
```

The function ode45 calls the function `debt_deficit_system` that is given by the following MATLAB® code.

The ODE system

```
function dd = debt_deficit_system(t,y,p)

   %Parameters
   b = p(1);
   r = p(2);
   g = p(3);

   %Variables
   d = y(1);

   %The ODE model
   dd = b +(r-g)d;
end
```

Finally, we plot the time history of $d(t)$ and indicate the location of the equilibrium point through the following MATLAB® code.

Visualisation

```
%Equilibrium point
d_eq = b/(g-r); %Solve dd/dt = 0 => b+(r-g)d_eq = 0

%Plotting
figure(1)
hold on
plot(t,d,'k','LineWidth',2)
plot([t_start,t_end],[d_eq,d_eq],'k:','LineWidth',2)
xlabel('time (t)'),ylabel('d(t)')
hold off, box on
```

Experiment with the code

Using the above MATLAB® code, the reader should be able to reproduce Figures 6.1–6.4, and is also encouraged to experiment with different parameter values and initial conditions.

6.4 CONCLUDING REMARKS, EXTENSIONS AND CHALLENGES

In this chapter, we discussed the debt-deficit dynamics using a basic set-up, beginning with government budget identity (6.1). The question of sustainability of debt has always been a challenging question in fiscal policy deliberations, and more so in the context of the Covid-19 pandemic in 2020–2021. The stability condition (6.10) tells us that the solvency of public debt depends crucially on whether the growth rate of the economy is higher or lower than the real interest rate. The first scenario is if the growth rate is higher than the interest rate where the economy converges to a stable debt-GDP ratio and the question of solvency does not arise. However, in the second scenario, if the growth rate falls below the interest rate the question of solvency becomes critical in fiscal policy deliberations. Interestingly, the former scenario was characteristic of the advanced economies like the US and Germany in 1960s and 1970s and the latter seems to be the case in the recent decades.[2]

We saw in the analysis that the role of the primary surplus ($b < 0$) comes into play. This is particularly relevant in the unstable scenario, when the interest rate is greater than the growth rate, where a substantial primary surplus may be required to reduce the debt burden (see Figure 6.4). However, increasing a large primary surplus requires either painful expenditure cuts or increases in taxation. Both these avenues have adverse income distributional impacts and other supply-side effects all of which may not be economically feasible and political viable.

The model discussed in this chapter is limited in the sense that we have restricted ourselves to a linear model and the growth rate, interest rate and government's primary balance all are considered as exogenous parameters. However, a more general model that takes the feedback effects between these variables into account would be useful from a policy perspective. We can easily see this from the model presented here. Consider the stable case where the growth rate exceeds the real interest rate, and where a large primary deficit is associated with a high, but stable debt-GDP ratio. As long as the increasing levels of debt induce growth via increasing the investment in new capital stock and increasing employment and are higher than the interest rate, the increasing levels of debt must be sustainable. However, if the higher growth is mainly driven by consumption and asset price increases then the increasing debt-GDP ratio may seem to be unsustainable. Further, the perceived risk associated with asset-price driven growth can become unsustainable in the eyes of the investors, which may result in higher risk premiums and higher interest rates on new borrowing for the government. In the absence of such non-linear feedback between the variables, the linear articulation limits our understanding of the debt-deficit dynamics.

NOTES

1 We have adapted the standard textbook exposition of modelling the debt-deficit dynamics as in [1].

2 See Carlin and Soskice [1], pp. 525–526.

BIBLIOGRAPHY

[1] Carlin, W. and Soskice, D.W. (2015), *Macroeconomics: Institutions, Instability, and the Financial System,* Oxford University Press, US.

[2] Reinhart, C.M. and Rogoff, K.S. (2010), Growth in a Time of Debt, *American Economic Review,* Vol. 100(2), pp. 573–78.

7 EXPECTATION DYNAMICS AND HYPERINFLATION

In Chapter 4, we studied a simple Keynesian model of macroeconomic income determination. The economy was viewed as an 'aggregate' entity and was analysed using the aggregate income-expenditure relation. While the model helped us to derive the equilibrium level of output, there are couple of questions that were not addressed. First, there is no story of what happens to prices (and inflation) in the model. Second, the aggregate analysis does not tell us how individual agents adapt their behaviour for any policy changes. For instance, an increase in taxes can impact on both the current and future spending plans of consumers and investors. Therefore, policy changes impact not only on current decisions but also future decisions by impacting on individuals' expectations about the future. This is important from a policy evaluation perspective. Policy makers in governments and central banks need to consider the impact their policy stance may have on individual agents' expectations because of its effect on current and future spending patterns. In the aggregate Keynesian models, studied in Chapters 4 and 5, economic policy analysis and predictions do not take into account the adaptive behaviour of individual agents, which in fact can impact the equilibrium value of the system. Therefore, aggregate macroeconomic models need to be *micro-founded,* so that they take into account how some of the behavioural parameters, so called 'deep parameters' of the system evolve with respect to policy changes. This is known as the *Lucas critique* of econometric policy evaluation and the Cagan's model, discussed in this chapter, provides a simple and intuitive way to illustrate this point.

7.1 ECONOMIC PROBLEM

In this chapter we will introduce a seminal study in *monetary economics* published in 1956 by Philip Cagan [1], an economist at the National Bureau of Economic Research (NBER) in the USA. The main aim of Cagan's paper

DOI: 10.4324/9780429324406-7

was to explain episodes of hyperinflation, which is defined as a period in which prices rose by more than 50 percent a month. For instance, during 1920–23 the average monthly inflation in Germany was 322 percent, whereas in Poland it was 81.4 percent. Cagan presented a formal model of hyperinflation episodes in six European countries after World War I and Hungary and China after World War II. The empirical results appeared to confirm the *quantity theory of money* hypothesis that hyperinflation is caused by the growth in nominal money stock. The model can also be used to describe the deflationary situation that was witnessed during the Great Depression in the 1930s.

Quantity theory of money

The underpinning logic in Cagan's model is driven by the *quantity theory of money,* which is given by $MV = PY$, where M is the quantity of money supplied by the Central Bank, V is the 'velocity' of money circulation in the economy, i.e. how many times money changes hands in a certain time, which is determined by the payment technology like debit/credit cards etc., P is the general price level and Y is the full-employment level of output. The theory says that given V and Y, any increase in money supply will lead to a rise in the price level.

Although set in the quantity theory of money framework, the interesting point about Cagen's model is to highlight the dynamic interaction between macroeconomic level monetary policy decisions and micro-level behaviour of individual agents in the study of inflation. While the equilibrium of the model is determined by the intersection of money demand and money supply, the dynamics of adjustment towards the equilibrium is driven by the interaction between the aggregate money-supply decisions at the macro level and the micro level individual agents' expectations about future money supplies. While Cagan's model used *adaptive expectations,* where expectations about the future are learned from the difference between past expectations and actual realisations, we will analyse the model using few simpler assumptions on agents' expectations and analyse the dynamics of the system. For instance, we explore the special case of 'perfect foresight', where agents' expectations coincide with the actual outcome. In the deterministic setting, the perfect foresight model will coincide with the scenario of *rational expectations* and we will discuss the possibility of emergence of the so-called *rational bubbles* in this case.

7.2 MODELLING

Cagan assumed that the demand for money balances M_t^d, in a period t, is negatively related to the expected rate of price inflation π_{t+1}^e, in the following period, as given by the *money demand function*

$$M_t^d = e^{-\alpha\pi_{t+1}^e}, \qquad \alpha > 0, \tag{7.1}$$

where the parameter α captures the responsiveness of money demand with respect to the expected inflation. The money demand function (7.1) implies that if agents expect inflation to increase in the next time period, then they know that the real value of their money balances will be diminished and hence they will desire to hold less money. Viewed at time t, the expected inflation is defined as the expected growth rate of the price in period t to $t + 1$ and given by

$$\pi_{t+1}^e = \frac{p_{t+1}^e - p_t}{p_t}. \tag{7.2}$$

The money supply M_t^s is assumed to be controlled by the central bank and is exogenously given in the model and thus the *real money supply* is given by

$$\frac{M_t^s}{P}. \tag{7.3}$$

The money market is in equilibrium when the real money demand equals the real money supply and from (7.1) and (7.3) we get

$$\frac{M_t^s}{P_t} = e^{-\alpha\pi_{t+1}^e}. \tag{7.4}$$

The economic meaning of the equilibrium condition (7.4) is the same as the money demand function (7.2). When agents expect higher inflation tomorrow, they will reduce their money balances as their real value is diminished under higher inflation.

We can now simplify the analysis by linearising the equilibrium condition (7.4). We first take the natural logarithm of (7.4) and get

$$\ln(M_t) - \ln(P_t) = -\alpha\pi_{t+1}^e. \tag{7.5}$$

Next we approximate (7.2) to linear order through

$$\pi_{t+1}^e = \frac{P_{t+1}^e - P_t}{P_t} \approx \ln\left(1 + \frac{P_{t+1}^e - P_t}{P_t}\right) = \ln\left(\frac{P_{t+1}^e}{P_t}\right). \tag{7.6}$$

The equilibrium condition (7.5) now becomes

$$\ln(M_t) - \ln(P_t) = -\alpha\left(\ln\left(P_{t+1}^e\right) - \ln(P_t)\right). \tag{7.7}$$

and by denoting

$$m_t = \ln(M_t), p_t = \ln(P_t) \text{ and } p_{t+1}^e = \ln\left(p_{t+1}^e\right)$$

we can rewrite (7.7) as

$$m_t - p_t = -\alpha\left(p_{t+1}^e - p_t\right), \tag{7.8}$$

where m, p and p^e are natural logarithms of the nominal money supply, general price level and expected price level, respectively. Collecting the price terms p_t in (7.8) we get

$$p_t = \frac{\alpha}{1+\alpha} p_{t+1}^e + \frac{1}{1+\alpha} m_t. \tag{7.9}$$

Equation (7.9) is the dynamical system that describes the evolution of price in this model. In economic terms, (7.9) shows that today's price p_t is determined by current money supply m_t and the individual agent's expectation about tomorrow's price level p_{t+1}^e. However, without specifying how expectations are formed, i.e., by defining the p_{t+1}^e term, we cannot determine the price p_t. In the following section we consider three different models of expectations formation by agents and analyse the price dynamics under these models. The expectation models analysed here are:

1. **Naïve expectation**, where

$$p_{t+1}^e - p_t = \gamma(p_t - p_{t-1})$$

for some constant $\gamma > 0$.

2. **Perfect foresight**, where

$$p_{t+1}^e = p_{t+1}.$$

3. **Rational expectation**, where

$$p_{t+1}^e = E_t[p_{t+1}]$$

with the term $E_t[P_{t+1}]$ being the expected value of P_{t+1}.

7.3 ANALYSIS, SIMULATION AND VISUALISATION

7.3.1 THE NAÏVE EXPECTATION MODEL

In the first instance, we assume that agents form their expectations about the future in a naïve manner where they expect the future rate of inflation to hover just above the historical trend and given by

$$p_{t+1}^{e} - p_t = \gamma\left(p_t - p_{t-1}\right), \quad \gamma > 0. \tag{7.10}$$

The parameter γ is a 'correction' parameter that incorporates the deviation of the current price level with that of the previous period's price. If the current price is above the previous period's price, the agent expects an upward momentum in the next period's price level. On the other hand, if the current price is below the previous period's price, the agent expects a downward momentum in the next period's price. Therefore, the agent is mechanically following the trend, which we call here the 'naïve' expectation behaviour.

Substituting the expectation rule (7.10) in the price-dynamics equation (7.9) and simplifying yields the first-order difference equation

$$p_t = -\frac{\alpha\gamma}{1-\alpha\gamma} p_{t-1} + \frac{m_t}{1-\alpha\gamma}. \tag{7.11}$$

We assume that money supply is exogenously advanced by the central bank and further suppose that the agents believe that the central bank is committed to keeping the nominal money supply in the economy constant, say $m_t = \bar{m}$, we can solve the first-order different equation and study its dynamic properties. With the assumption of constant money supply, the difference equation (7.11) becomes

$$p_t = -\frac{\alpha\gamma}{1-\alpha\gamma} p_{t-1} + \frac{\bar{m}}{1-\alpha\gamma} \tag{7.12}$$

and the solution to this difference equation is

$$p_t = \left(p_0 - p^*\right)\left(\frac{-\alpha\gamma}{1-\alpha\gamma}\right)^t + p^*, \tag{7.13}$$

where the equilibrium price level is given by $p^* = \bar{m}$. In economic terms, this means that the equilibrium price level is determined by the given level of money supply.[1] The solution shows that starting from any initial arbitrary value of price level p_0 the system will converge to the equilibrium price level p^* if the condition

$$\left|\frac{-\alpha\gamma}{1-\alpha\gamma}\right| < 1 \Leftrightarrow \alpha\gamma < |1-\alpha\gamma| \Rightarrow 0 < \alpha\gamma < \frac{1}{2} \tag{7.14}$$

holds, where we have $\alpha\gamma > 0$. This is the stability condition for the price dynamics equilibrium, i.e. the price will converge to the equilibrium $p^* = \overline{m}$ only when condition (7.14) is satisfied; otherwise the prices will diverge from the equilibrium and the equilibrium point is said to be unstable, meaning that there will be *inflation* in the economy. Figure 7.1 shows both stable and unstable cases.

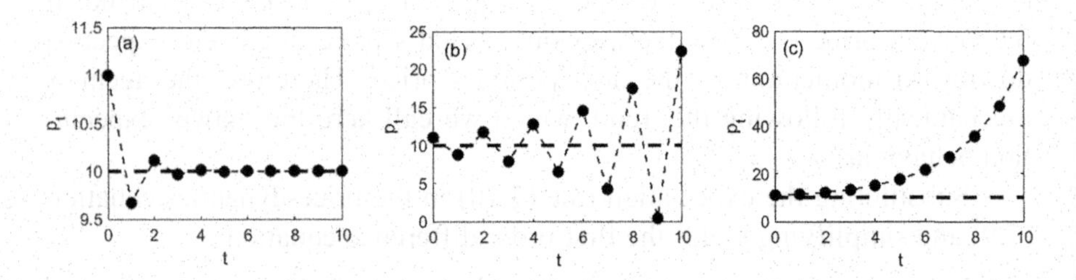

Figure 7.1: Three examples of dynamics given by (7.12) with constant money supply $\overline{m} = 10$ (and thus $p^* = 10$) and initial condition $p_0 = 11$. In (a) $\alpha = 0.5$, $\gamma = 0.5$ ($\alpha\gamma = 0.25$), which give a oscillatory dynamics about the stable equilibrium point, in (b) $\alpha = 0.8$, $\gamma = 0.8$ ($\alpha\gamma = 0.64$), which yields an oscillatory dynamics about the unstable equilibrium point, and in (c) $\alpha = 4$, $\gamma = 0.8$ ($\alpha\gamma = 3.20$), which results in a monotonic dynamics about the unstable equilibrium point.

The stable case is shown in Figure 7.1(a) and the unstable cases are shown in Figures 7.1 (b) and (c). It is clear from the figures that small changes in the parameters, both α and γ, change the stability of the system, which highlights how sensitive the system dynamics is to the agents' behaviour. The parameter α captures how agents' money demand is responsive to expected inflation and γ represents agents' belief on how prices are going to evolve in the future. Figure 7.1 shows that the model yields persistent inflation when the responsiveness of money demand to expected inflation (α) increases even when the correction parameter (γ) remains the same (compare Figures 7.1(b) and (c)). In a way, the above analysis shows that persistent inflation can happen purely through the agents' expectation behaviour in a model that assumes inflation to be a monetary phenomena, as postulated by the quantity theory of money. We can now explore the system along this line and verify the hypothesis under different types of expectation-formation behaviour.

7.3.2 THE PERFECT FORESIGHT MODEL

As an alternative to the naïve expectation case, let us suppose that the agents have *perfect foresight,* where the expected values coincide with the actual materialised values so that

$$p_{t+1}^{e} = p_{t+1}. \tag{7.15}$$

Substituting the rule (7.15) into (7.9) and rearranging yields the first-order difference equation

$$p_{t+1} = \frac{1+\alpha}{\alpha} p_t - \frac{1}{\alpha} m_t \tag{7.16}$$

for the price evolution. Assuming a constant money supply, as in Section 7.3.1, where $m_t = \bar{m}$, we again see that the equilibrium price $p^* = \bar{m}$. If the initial price p_0 equals the equilibrium price p^* then the system will stay at this price forever. However, in order to see what happens if the initial price differs from the equilibrium price, we solve the differential equation (7.16) and get

$$p(t) = \left(p_0 - p^*\right)\left(\frac{1+\alpha}{\alpha}\right)^t + p^*, \tag{7.17}$$

with $p^* = \bar{m}$. We immediately note that the price will always diverge monotonically away from p^* since

$$\frac{1+\alpha}{\alpha} > 1 \tag{7.18}$$

for all $\alpha > 0$. This implies that for all initial conditions, except the one that coincides with the equilibrium price, the price monotonically diverges away from the equilibrium value, and the equilibrium is thus unstable, and this divergent dynamics is driven purely by agents' perfect foresight rather than by the growth in money supply.

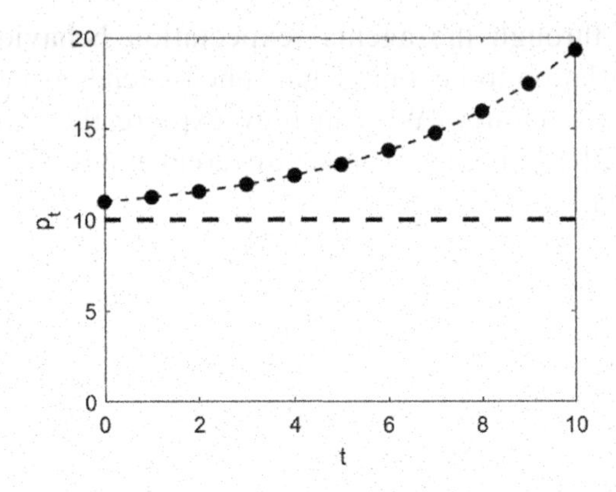

Figure 7.2: A solution to (7.16) for $m_t = \overline{m} = 10$, $\alpha = 4$ and $p_0 = 11$.

In Figure 7.2 we see a typical case of the divergent dynamics when agents have perfect foresight and it is often referred to as a case of a *rational bubble* in the literature [2], and in this deterministic setting the perfect foresight assumption is equivalent to the assumption of *rational expectations.* In order to understand the intuition behind the divergent dynamics in this setting, consider the case where the price level is growing at an increasing rate, so that the expected inflation rate is rising over time.[2] In terms of the money market equilibrium, an increase in price level implies that the demand for money, and the real money supply, would diminish. However, depending on the relative difference between the real money supply and money demand (see (7.4)) the model can result in hyperinflation (or deflation). If the fall in money demand is greater than the fall in the money supply, then the actual price level will rise, leading to a rise in the expected price. This is the case where the parameter α is large, which results in a steeper downward facing money demand curve and makes the price diverge monotonically away from the equilibrium price. Therefore, higher actual price leads to higher expected price which in turn leads to further increase in actual price level. Thus continued inflation, or hyperinflation, is fuelled by self-fulfilling expectations. Hyperinflation in this case (or deflation in the opposite case of α being a smaller value) is driven by the self fulfilling expectations as opposed to central bank expanding (contracting) money supply at an increasing rate.

There are many objections to the rational bubbles argument in the literature [3]. The assumption that agents expect that the price can rise indefinitely, in the absence of money growth, is highly problematic. On the other hand, if agents do not believe that prices can rise indefinitely, then the only solution for the system is where the system's initial condition coincides with the steady-state value and the system is stuck at the steady state value forever, which is also unrealistic. Thus, skeptics argue that these rational bubbles are unreasonable characterisations of real-world inflation dynamics.

The scenarios we have seen so far represent two extreme cases of expectation formation where agents' expectations, at any given level of money supply, drive the dynamics of the model. However, the fundamental point about the quantity theory of money, upon which Cagan's model rests, is the role money-supply growth has in driving inflation. Therefore, in the case where money supply growth is varying, it is clear that agents' expectation about the future price level must be based on the future money-supply growth. Let us next study a scenario where agents' expectation about future prices is ultimately derived from their expectation about the future money supply.

7.3.3 ANALYSIS OF THE RATIONAL EXPECTATION MODEL

In this scenario we assume that expectations are formed on the basis of rational scientific calculation of the economy, i.e. based on *rational expectations*. Unlike in the adaptive expectations scenario where agents learn from past deviations between the actual and expected values, the rational expectations set-up is a forward-looking iterative method where the future price is formally expressed as

$$p_{t+1}^{e} = E_t\left[p_{t+1}\right]. \tag{7.19}$$

In order to determine the price at each time, let us rewrite the money market equilibrium condition (7.9) using (7.19) so that

$$p_t = \frac{\alpha}{1+\alpha} E_t\left[p_{t+1}\right] + \frac{1}{1+\alpha} m_t \tag{7.20}$$

at time t, and

$$p_{t+1} = \frac{\alpha}{1+\alpha} E_{t+1}\left[p_{t+2}\right] + \frac{1}{1+\alpha} m_{t+1} \tag{7.21}$$

at time $t+1$, from which we get

$$E_t\left[p_{t+1}\right] = \frac{\alpha}{1+\alpha} E_t\left[E_{t+1}\left[p_{t+2}\right]\right] + \frac{1}{1+\alpha} E_t\left[m_{t+1}\right] \tag{7.22}$$

or equivalently

$$E_t\left[p_{t+1}\right]=\frac{\alpha}{1+\alpha}E_t\left[p_{t+2}\right]+\frac{1}{1+\alpha}E_t\left[m_{t+1}\right], \tag{7.23}$$

where we use the *law of iterated expectations* rule that says

$$E_t\left[E_{t+1}\left[p_{t+2}\right]\right]=E_t\left[p_{t+2}\right]. \tag{7.24}$$

The law of iterated expectation implies that an agent's expectation today about tomorrow's expectation of the day after tomorrow's price is equal to today's expectation of the day after tomorrow's price. A simpler way to understand the intuition behind the law is to imagine a forward contract that a supplier places for supplying an agreed quantity of an agricultural commodity, say wheat, at an agreed price in a month from now. Then the law of iterated expectation implies that the future price that is set today is an average of expectation made today about tomorrow's expectation of the day after tomorrow's expectation and so on, of the end of the period price.

If we continue the iteration process for p_{t+2}, p_{t+3}, … and substitute for $E_t\,[p_{t+2}]$, $E_t\,[p_{t+3}]$, … in the price equation (7.20) we get

$$p_t=\frac{1}{1+\alpha}m_t+\frac{1}{1+\alpha}\frac{\alpha}{1+\alpha}E_t\left[m_{t+1}\right]+\frac{1}{1+\alpha}\left(\frac{\alpha}{1+\alpha}\right)^2E_t\left[m_{t+2}\right]+\ldots \tag{7.25}$$

or

$$p_t=\frac{1}{1+\alpha}\sum_{k=0}^{\infty}\left(\frac{\alpha}{1+\alpha}\right)^kE_t\left[m_{t+k}\right], \tag{7.26}$$

with $E_t\,[m_t] = m_t$. The expression (7.26) simply shows that today's price level p_t is a weighted sum of today's and the expected future nominal money supply, where the weights get smaller the further in the future we look. Thus, in the scenario where the money supply growth is a variable, the rational expectation calculation about the future price level converges to agents' expectation about the future money supply, and therefore the equilibrium price is determined by the money supply.

The rational expectation hypothesis

The reader might wonder about the wisdom of the rational agent to know the underlying model of the economy, i.e. the quantity theory of money, and look at the money supply process for estimating expected inflation. This is one of the fundamental assumptions of the rational expectations hypothesis. It assumes that the rational agent knows the 'correct model' of the economy. Even if the agent makes some error in their expectation calculation, the rational expectations hypothesis assumes that the errors are uncorrelated over time and tends to zero on average. Furthermore, the hypothesis also assumes that the errors are uncorrelated across the variables in the model. Readers might wonder, given the power of rationality that agents seem to posses, how errors appear in the first place. We will discuss this point below.

We can first examine the rational expectations expression (7.26) with different money supply assumptions. Suppose that the central bank is committed to keeping the nominal money stock in the economy constant, such that $m_t = \bar{m}$. As we discussed in the naïve expectation scenario, the equilibrium price level will be exactly equal to $\bar{m}$. We can see this, using the rational expectation solution (7.26), as

$$\begin{aligned} p_t &= \frac{1}{1+\alpha}\sum_{k=0}^{\infty}\left(\frac{\alpha}{1+\alpha}\right)^k E_t\left(m_{t+k}\right) = \frac{\bar{m}}{1+\alpha}\sum_{k=0}^{\infty}\left(\frac{\alpha}{1+\alpha}\right)^k \\ &= \frac{\bar{m}}{1+\alpha}\frac{1}{1-\frac{\alpha}{1+\alpha}} = \frac{\bar{m}}{1+\alpha}(1+\alpha) = \bar{m}. \end{aligned} \tag{7.27}$$

Therefore, when the money supply is constant the price level is constant over time.

Second, we instead consider the case where the central bank conducts its monetary policy based on discretion without any commitment to a particular level of money supply. In such a situation, assume that the rational agents' best guess is to believe that the money supply process follow a stochastic process such as

$$m_t = \bar{m} + \varepsilon_t, \tag{7.28}$$

where ε_t is a random variable with zero mean and represents the surprise shock to money supply at time t. In economic terms this means that there is a discretionary increase money supply over and above the constant part $\bar{m}$, say, some target base level of money supply. Furthermore, assume that the effect of the random component, the 'shock' component, lasts only one period, and the effect of the shock remains temporary. In policy terms this means that the central bank wants to keep the money supply at $\bar{m}$ and whenever the money supply is discretionarily moved away from this level, the central bank's action to bring the money supply back to $\bar{m}$ impacts the economy with a one-period lag.

Now, substituting the stochastic money supply rule (7.28) in the rational expectations equilibrium price (7.26) yields

$$p_t = \frac{\bar{m}}{1+\alpha}\sum_{k=0}^{\infty}\left(\frac{\alpha}{1+\alpha}\right)^k E_t\left[m_{t+k}\right], \tag{7.29}$$

and since

$$E_t\left[m_t\right] = E_t\left[\bar{m}+\varepsilon_t\right] = \bar{m}+\varepsilon_t \quad \text{and} \quad E_t\left[m_{t+k}\right] = E_t\left[\bar{m}+\varepsilon_{t+k}\right] = \bar{m}+0 = \bar{m}$$

for k = 1, 2, …, the equilibrium price level follows the stochastic process

$$p_t = \frac{1}{1+\alpha}\varepsilon_t + \bar{m}, \tag{7.30}$$

which is essentially a random walk about $p = \bar{m}$. An example of this process can be seen in Figure 7.3, where the random variable ε_t is taken from a normal distribution with mean 0 and variance 1. Recall that the parameter $\alpha > 0$ represents the responsiveness of money demand with respect to expected inflation. Further note that the money supply shocks, or the discretionary

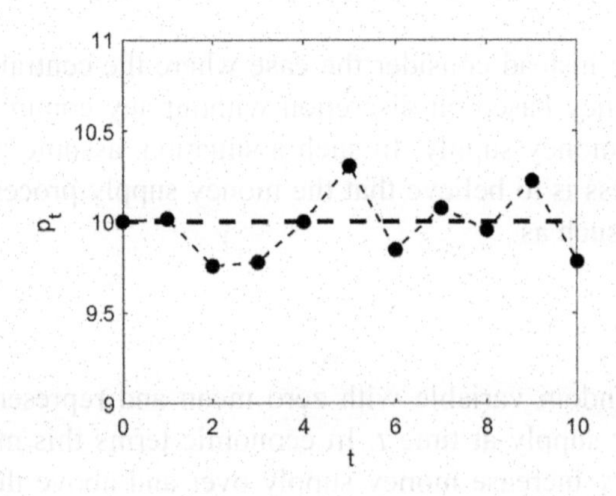

Figure 7.3: A solution to (7.30) for $\bar{m}$= 10, α = 4, p_0 = 10 and $\varepsilon \sim N(0,1)$.

increases in money supply, will remain temporary since $\frac{1}{1+\alpha}<1$. This simply means that the rational agents are responding to the money supply shocks by reducing their money demand in anticipation of inflation. Therefore, the price level today p_t does not increase to the full extent of the shock because the rational agent expects the inflation to be temporary and that the price level will return to the level corresponding to the $\bar{m}$ level of money supply.

However, the central bank looking at aggregate money supply and inflation data might interpret the less than proportionate increase in inflation as providing it with more wiggle room for further expansion of the money supply. The discrepancy between the actions of rational agents and the central bank cannot go on and will end as soon as the former begin to realise that the latter is increasing the target base level rather than the discretionary increases in the money supply via the temporary shocks. Once the discrepancy is resolved, inflation will be tracked correctly to the extent of the increase in the money supply.

The preceding example provides an illustration of the so called *Lucas critique* of policy evaluation: a change in the macroeconomic policy will change not only the policy variable, money supply in this example, but also rational agents' expectations. The impact of policy plays out in the short run until rational agents catch up with the intention of the central bank that is obscured by its discretionary actions. Therefore, central banks' policy evaluation, using the aggregate macroeconomic data without taking into account the reaction of rational agents, can render misleading analysis of the effect of a policy change. This is the economic essence of the critique.

The Lucas critique

Robert Lucas critiqued the then prevailing large-scale macro-econometric models that lacked 'microfoundations' in that they do not specify how economic agents' preferences change with policy changes.

> Given the structure of an econometric model consists of optimal decision rules of economic agents, and that optimal decision rules vary systematically with changes in the structure of series relevant to the decision maker, it follows that any change in policy will systematically alter the structure of econometric models.
>
> (Lucas, R. E. [4], p. 41)

The critique paved the way for the modern micro-foundations approach to macroeconomics where the evaluation of agent's behavioural parameters, or the 'deep parameters', in various policy scenarios are one of the central considerations.

The implications of the Lucas critique are overarching in the sense of the following arguments. The discretionary way in which the central bank conducts its policy making can thwart rational agents' expectations by introducing uncertainties in their expectations calculations. By extension, the critique also applies to government's fiscal policy making, where discretionary spending or taxation decisions can distort rational agents' expectations and their inter-temporal consumption and investment decisions. Therefore, the force of the Lucas critique is to argue that in a world where central banks and governments conduct a "rule-based" policy making, the scope for expectational errors by the rational agents is lower and consequently their inter-temporal decisions would be more certain thus leading to smaller fluctuations in the economy. The implications of the Lucas critique were profound in a way that it has influenced macroeconomic theory as well as applied macroeconomic policy making since the 1980s.

7.3.4 SIMULATION AND VISUALISATION

To simulate the price evolution for the three cases of price expectation in Sections 7.3.1–7.3.3, as seen in Figures 7.1–7.3, we can use very similar MATLAB® codes as in Sections 2.3 and 4.3 since they all solve first-order difference equations. In all three cases we start by clearing the memory and closing all windows, by writing

Clear variables and close windows

```
%Preamble
clear all % Clears all variables
close all % Closes all figure windows
```

The code for simulating the price evolution using the naïve expectation model analysed in Section 7.3.1 and plotted in Figure 7.1(a) is given in the following.

Initial conditions and naïve expectation simulation

```
%Parameters
alpha = 0.5;
gamma = 0.5;
m     = 10;
```

```
%Intial Condition
p0 = 11;

%Simulation parameters
n= 10; % Number of iterations

%Simulation
p(1) = p0;
for k = 1:n
    p(k+1) = -alpha*gamma/(1-alpha*gamma)*p(k)+...
             m/(1-alpha*gamma);
end
```

The code for simulating the price evolution using the perfect foresight model analysed in Section 7.3.2 and plotted in Figure 7.2 is given as follows.

Initial conditions and perfect foresight simulation

```
%Parameters
alpha = 4;
m     = 10;

%Intial Condition
p0 = 11;

%Simulation parameters
n = 10; % Number of iterations

%Simulation
p(1) = p0;
for k = 1:n
    p(k+1) = (1+alpha)/alpha*p(k) - m/alpha;
end
```

The code for simulating the price evolution using the rational expectation model analysed in Section 7.3.3 and plotted in Figure 7.3 is given as follows.

Initial conditions and rational expectation simulation

```
%Parameters
alpha = 4;
m = 10;
```

```
%Intial Condition
p0 = 10;

%Simulation parameters
n = 10; % Number of iterations

%Simulation
p(1) = p0;
for k = 1:n
   e = randn(1,1);
   p(k+1) = 1/(1+alpha)*e + m;
end
```

Finally, we can plot the time histories of the price p_t for each of the three cases introduced above and also indicate the location of the fixed point through the following MATLAB® code.

Visualisation

```
%Fixed point
pfp = m;

%Plotting
figure(1),hold on
plot(0:n,p,'ko','MarkerFaceColor','k')
plot([0,n],[pfp,pfp],'k--','LineWidth',2)
xlabel('t'),ylabel('p_t')
box on
hold off
```

7.4 CONCLUDING REMARKS, EXTENSIONS AND CHALLENGES

Cagan's model was designed to explain hyperinflationary episodes. His model was set in the framework of quantity theory of money, where there is a strict dichotomy between the monetary and real sectors of the economy and the analysis of hyperinflation as a purely monetary phenomenon. The monetary equilibrium is driven by the supply and demand for money, where the former is exogenously supplied by the central bank and the latter is a function of the expected rate of inflation. The novelty of the model is that it shows the possibility of hyperinflation in this standard set up via agents' expectation formation. In his model, Cagan employed the so called adaptive expectation, where expectations of the future are formed by applying exponentially declining weights to past inflation rates.

The aim of this chapter is to study the dynamics of Cagan's model using a few different expectations hypotheses. For the purposes of exposition, we started with a simpler expectation formation rule called the 'naïve' adaptive approach to study the stability of the equilibrium. The analysis showed how sensitive the stability of the system is with respect to the agents' behavioural parameters, namely, the responsive of money demand to expected inflation (through α) and the correction parameter γ. As we saw in Figure 7.1, the model yields persistent inflation for certain values of the parameters.

In the second model, we studied the case of 'perfect foresight'. Interestingly, the stability analysis showed the possibility of rational bubbles and the hyperinflation in this case is driven purely by the self fulfilling expectations at any given money supply. This may sound unrealistic, particularly for models in the framework of the quantity theory of money. A more generalised model would encompass perfect foresight as a special case and the rational expectations approach, discussed in Section 7.3.3, provides such a model.

The rational expectations model was analysed in two scenarios – first, with a given level of money supply and, second, in the case where the central bank discretely changes the money supply around a base level. The analysis showed the dynamics is driven by agents' expectation about the size of the impact of the money supply shocks on prices. If the impact is less than proportionate, rational agents will see the shock as temporary and inflation will track the level of money supply in the next time period. In this case, the central banks could use a less than proportionate increase in inflation as a signal and the freedom to push the money supply further. This may create a discrepancy between rational agents' expectation calculation and actual money supply growth in the economy, and consequently leading to fluctuations in the macroeconomic variables. However, once the rational agents realise that the central bank's actions have led to permanent increase of money supply above the base level, the inflation will fall in line with the increased money supply growth.

Cagan's model also provides a simple example to discuss how the stability of the system is driven by the interaction between macro-level policy, such as money supply, and microeconomic agents' behaviour. In that sense, it provides an example of the Lucas critique that we discussed above. The rational expectations model brings out the dynamics, albeit in a limited way. The implications of the critique were profound in that it replaced the 'structural' approach with the 'micro-foundational' approach to the study of macroeconomics.

The question that arises in this context is how to describe the fluctuations in output, employment and inflation in the rational expectations framework, where agents have knowledge of the underlying model of the economy. As partly discussed in Section 7.3.3, fluctuations in macroeconomic variables arise due to unanticipated shocks. These shocks could be driven by the discretionary policy stance of the central banks or governments, or due to various institutional frictions, or due to natural phenomena like epidemics. The study of fluctuations, or 'business cycles', is one of the central themes of macroeconomics and we will study this question using a micro-founded macroeconomic model in Chapter 12.

NOTES

1 Note from the solution that for certain values of α and γ we get $\alpha\gamma = 1$ and where the model (7.12) is not mathematically defined. The relation, between α and γ, in fact, stipulates the range of values $\gamma < 1$ and $\alpha > 0$ for which the model yields meaningful results.

2 Think of comparative statics analysis performed on the conventional downward sloping money demand curve and the vertical money supply curve.

BIBLIOGRAPHY

[1] Cagan, P. (1956), The Monetary Dynamics of Hyperinflation, In M. Friedman ed., *Studies in the Quantity Theory of Money,* pp. 25–117, Chicago: University of Chicago Press.

[2] Diba, B.T. and Grossman, H.I. (1988), The Theory of Rational Bubbles in Stock Prices, *The Economic Journal,* Vol. 98, No. 392, pp. 746–754.

[3] DeRosa, D.F. (2021), *Bursting the Bubble: Rationality in a Seemingly Irrational Market*, CFA Institute Research Foundation.

[4] Lucas, R.E. (1976), Econometric Policy Evaluation: A Critique, In Brunner, K. and Metzler, A. (eds) *The Phillips Curve and Labour Markets Carnegie-Rochester Conference Series on Public Policy.* Vol. 1, New York: American Elsevier, pp. 19–46.

8 THE DORNBUSCH EXCHANGE RATE OVERSHOOTING MODEL

In Chapter 7 we studied how the dynamics between macro-level monetary policy and individual agents' expectations play a central role in understanding inflation and different hyperinflationary scenarios. In this chapter, we extend the analysis further to study the complex dynamics of exchange rates in an open economy context. Here we consider one of the most influential papers in the field of international economics: *Expectations and exchange rate dynamics* by Rudiger Dornbusch published in the Journal of Political Economy in 1976 [1]. The paper was written in the context when the developed countries in the western world had just made the transition from fixed to flexible exchange rates. Policy makers and policy analysts were grappling with the volatility in the exchange-rate markets as exchange-rate changes did not mirror international inflation differentials between the trading countries. Dornbusch's paper was timely and quite revealing in that it showed that the exchange-rate volatility is not necessarily due to the myopia or herd behaviour of traders as a response to adverse monetary shocks, but rather it is inherent to the system due to the different speed adjustment of prices in the economy. The model relies on what is known as the *overshooting phenomenon* to explain the volatility in the exchange rates. While his conjecture about why exchange rates overshoot may not be an empirical success, the idea of the differential speed of changes in prices in various markets – the sluggish adjustment of goods market prices vis-a-vis the instantaneous adjustment in the currency markets prices – was certainly more realistic and useful from a policy perspective.

8.1 ECONOMIC PROBLEM

We shall briefly outline the model and discuss some important concepts. There are two core equations in the model, the *uncovered interest rate parity* (UIP) condition and the *money demand* function. The

DOI: 10.4324/9780429324406-8

uncovered interest rate parity condition states that the home (nominal) *interest rate* on bonds i (for example, government treasury bonds) must equal the foreign interest rate i^*, which is assumed to be constant here, plus the expected rate of depreciation of exchange rate e. From this we have that the home interest rate at time $t + 1$ is given by

$$i_{t+1} = i^* + E_t\,[e_{t+1} - e_t], \tag{8.1}$$

where we note that $E_t\,[\cdot]$ denotes agent expectations based on information at time t. The condition states that the interest gain from holding home-currency assets equals the loss of expected depreciation of home currency vis-a-vis the foreign currency. The following example will help us understanding the UIP condition.

Example: the UIP condition

Suppose we consider a European country, in the European monetary union, as the home country and the USA as the foreign country.

Let us first define the concept of the exchange rate, both the nominal and real exchange rates. The *nominal exchange rate (e)* of the home country is the amount N of home currency that can be bought with one unit of foreign currency, i.e. here we get

$$e = \frac{€N}{\$1}.$$

An increase in e gives that one US dollar can buy more Euros, which means that the Euro has depreciated (or weakened).[a] The nominal exchange rate thus shows the rate at which the currencies are traded.

To measure the rate at which home and foreign goods and services are traded, we define the *real exchange rate Q*, which also takes into account the relative price level between the two countries, and is defined as

$$Q = \frac{p^* e}{p},$$

where p^*e is the price of foreign goods expressed in home currency (in Euros) and p is the price of home goods. In a way Q is a measure of price competitiveness between the two countries. When the price of home goods falls relative to the price of foreign goods Q increases implying depreciation of the real exchange rate, which makes home exports more competitive for the foreign countries and imports become more expensive for domestic consumers.

With these definitions, we are now in a position to understand the UIP condition. Suppose we start with a situation where the home interest rate and the foreign interest rate are equal, say 2%, and that the home central bank increases its interest rate to 4% for controlling the domestic inflation. The positive interest rate differential between the home and foreign interest rates creates an arbitrage opportunity, which translates to an increased demand for Euro bonds. What happens to the exchange rate e in this situation? The interest differential, in favour of Euro bonds, results in a stronger demand for Euro currency, which forces the currency to appreciate vis-a-via the dollar. Since investors try to maximise their returns, the Euro will appreciate by 2% as soon as the interest rate decision is announced to eliminate the arbitrary opportunity. Assuming that the interest rate will be reversed to its initial level in the next period (or in the near future), the UIP condition states that the expected depreciation of the Euro over this period is equal to the interest gain from holding the higher yielding Euro bonds over the same period. This argument applies in the opposite situation as well.

[a] Since the nominal exchange rate is defined as the amount of home currency per one unit of foreign currency, a depreciation of the home currency is the same as an increase in the exchange rate e.

The second core equation of the model is the money demand function

$$M_t^d = f\left(i_t, y_t\right), \tag{8.2}$$

where the money demand M^d is negatively affected by the interest rate i, representing the speculative demand, and positively affected by the output y, arising from the transactions demand. A higher interest rate increases the opportunity cost of holding money and thereby reduces the demand for money. However, an increase in output, or income, raises the transaction demands for money. Assuming an exogenously given money supply M^s, the money market equilibrium condition is simply given when the supply of real money balance equals the demand for money, and is given by,

$$\frac{M_t^s}{p_t} = f\left(i_t, y_t\right). \tag{8.3}$$

In order to understand the intuition behind the exchange rate overshooting, we assume that the price level p_t is fixed, or moves sluggishly, and the output y_t is constant. Now, suppose that there is an unanticipated increase in the money supply. This means that the supply of real money balance will rise, and since the price level is fixed the demand for real money balance must rise to equilibrate the money market. However, given that the output is fixed, the only way the demand for money can increase is if the interest rate on domestic currency bonds falls. According to the UIP condition, the fall in domestic interest rate, relative to the given level of foreign interest rate, must trigger an upward adjustment in the exchange rate, i.e. the home currency depreciates, and an expected appreciation of the exchange rate over the duration of the interest rate differential. However, since rational agents, who know that money supply will increase the price level, will expect the currency to depreciate more than proportionally before it reverts to the original level. In fact, Dornbusch's model explains this behaviour where the exchange rate immediately depreciates more, or *overshoots*, than the long-run equilibrium value and then appreciates along with the increase in commodity prices. Thus, he showed that the exchange rate initially overshoots and then retracts to the long-run equilibrium value. As you will see from the analysis of the model the overshooting result is driven mainly by the relative speeds of adjustment in prices, i.e. the sluggish movement in the commodity prices and the instantaneous adjustment in the exchange rate.

8.2 MODELLING

Let us study the dynamics of the model in more detail.[1] The assumptions of the model are as follows: the economy is assumed to be a small open economy with perfect capital mobility (no controls on its capital account), where the exchange rate is fully flexible (market is determined) and assumed to adjust instantaneously for any 'news'. On the contrary, the commodity prices are assumed to be fixed, i.e. assumed to adjust slowly to their equilibrium value, and the economy's output is assumed to be at the full-employment level. At the micro level, agents are assumed to be rational and the model is set in the deterministic context.

To analyse the model, we first express the UIP condition (8.1) in continuous time so that

$$r(t) = r^* + \dot{e}(t), \tag{8.4}$$

where

$$r(t) = \ln(i(t)),\ r^* = \ln(i^*),\ e(t) = \ln(E) \text{ and } \dot{e}(t) = \frac{de}{dt} = \frac{d\ln(E)}{dt} = \frac{\dot{E}}{E}.$$

Note that under the rational expectations assumption, which in a deterministic setting is equivalent to perfect foresight, the expected exchange rate is equal to the actual exchange rate and thus $E[e] = e$.

Next, we focus on the money market equilibrium (8.3) and consider the money demand function

$$M^d = Y^{\phi} e^{-\alpha r}, \tag{8.5}$$

where $\alpha > 0$ represents the responsiveness of money demand to the interest rate and $\phi > 0$ represents the responsiveness of money demand with respect to income Y. Assuming the money supply M to be exogenous, the money market equilibrium can be written as

$$\frac{M}{P} = Y^{\phi} e^{-\alpha r}. \tag{8.6}$$

Taking the natural logarithm on both side of (8.6) and substituting the UIP condition (8.4), which incorporates agents' rational expectations, yields

$$p - m = -\phi y^F + \alpha r^* + \alpha \dot{e} \tag{8.7}$$

where

$$p = \ln(P),\ M = \ln(M) \text{ and } y^F = \ln(Y),$$

with y^F denoting the full-employment level of output. However, at equilibrium where all expectations are fulfilled, and both output and prices are at their steady-state values, the money market equilibrium condition becomes

$$\bar{p} - m = -\phi y^F + \alpha r^*. \tag{8.8}$$

Subtracting (8.8) from (8.7) gives rise to the dynamic equation for the exchange rate,

$$\dot{e} = \frac{1}{\alpha}(p - \bar{p}). \tag{8.9}$$

We can see that the exchange rate dynamics is determined by the deviation of the current price level from its equilibrium level. Note, the parameter α, which captures the responsiveness of money demand to interest rates, governs the speed of adjustment of prices in the exchange rate dynamics. We will return to this point later in Section 8.3.

8.2.1 COMMODITY PRICE DYNAMICS

Let us now model the price dynamics. The Dornbusch model assumes that output y ($= y^F$) is at the full-employment level and any excess demand (or excess supply) will lead to changes in the price level, or the economy adjusts through prices adjustment. The aggregate demand for domestic output depends on the relative price of domestic goods vis-a-vis foreign goods (that is $(e + p^*) - p$), the interest rate r, and the real income y.[2] For simplicity, we normalise the price of foreign goods, so that $p^* = 1$ and using these assumption, we can write the aggregate demand function as

$$\ln(D) \equiv d = v + \beta(e - p) + \gamma y - \sigma r, \tag{8.10}$$

where v is the intercept or shift parameter.

Since the economy is at full-employment level and any excess demand (or supply) adjusts only through prices, the price dynamics is given by

$$\dot{p} = \theta(d - y^F), \tag{8.11}$$

where $\theta > 0$ is the speed of adjustment of prices. Substituting the aggregate demand d from (8.10) in the price equation (8.11) yields

$$\dot{p} = \theta(v + \beta(e - p) + (\gamma - 1)y - \sigma r). \tag{8.12}$$

Thus, from (8.9) and (8.12) we have a two-dimensional system of ordinary differential equation given by

$$\dot{e} = \frac{1}{\alpha}(p - \bar{p}), \tag{8.13}$$

$$\dot{p} = \theta\beta(e - p) + \theta(v + (\gamma - 1)y - \sigma r), \tag{8.14}$$

where the former equation defines the exchange rate (e) dynamics, based on the UIP condition and the money-market equilibrium, and the latter equation defines the price (p) dynamics based on the goods-market equilibrium.

All equilibrium values are determined by setting $\dot{p}=0$ and $\dot{e}=0$. Since at equilibrium the expectations are fully realised, the domestic and foreign interest rates are equal, i.e. $r = r^*$, and using the equilibrium value for the price $\overline{p}$, we can determine the equilibrium value for exchange rate $\overline{e}$. Letting $\dot{p}=0$ and substituting the equilibrium for the interest rate (i.e, the money market clearing interest rate r^* obtained from (8.8)) and price level ($\overline{p}$) yields

$$\overline{e}=\overline{p}+\frac{1}{\beta}\left(\sigma r^{*}+(1-\gamma)y-v\right), \tag{8.15}$$

and the equilibrium of the two-dimensional system is thus given by $(\overline{e},\overline{p})$.

We now turn to the disequilibrium dynamics of the system. Given the $\dot{e}$ equation is expressed in terms of a price deviation from its equilibrium, it would be convenient if we express the $\dot{p}$ equation similarly.

To transform the price-dynamics equation (8.14) as deviations from the equilibrium, we first substitute the value of the interest rate r^*, the interest rate that yields the money market equilibrium (8.8), and further simplify as

$$\dot{p}=\theta\beta(e-p)+\theta\left(v+\frac{\sigma}{\alpha}(m-p)-\rho y^{F}\right), \tag{8.16}$$

where $\rho=\frac{\phi\sigma}{\alpha}+1-\gamma$.

Now subtracting the above equation (8.16) from the long-run equilibrium of the price dynamic equation ($\dot{p}=0$), with variables in their respective steady-state values given by

$$0=\theta\left(v+\beta(\overline{e}-\overline{p})+\frac{\sigma}{\alpha}(m-\overline{p})-\rho y^{F}\right), \tag{8.17}$$

results in the price dynamics equation with both the exchange rate and the price level expressed in terms of deviation from their respective long-run equilibrium values as

$$\dot{p}=-\theta\left[\left(\beta+\frac{\sigma}{\alpha}\right)(p-\overline{p})+\beta(e-\overline{e})\right]. \tag{8.18}$$

We now have the coupled system of exchange rate and price dynamics

$$\dot{e}=\frac{1}{\alpha}(p-\overline{p}), \tag{8.19}$$

$$\dot{p} = \theta\beta(e - \overline{e}) - \theta\left(\beta + \frac{\sigma}{\alpha}\right)(p - \overline{p}), \tag{8.20}$$

in the desired form where the variables are expressed as deviation from their respective long-run equilibrium values, which makes the economic interpretation of the analysis easier. We can now analyse the dynamics of the system.

8.3 ANALYSIS, SIMULATION AND VISUALISATION

Since the system (8.19)–(8.20) is linear, we can thus rewrite it as

$$\begin{aligned}\begin{pmatrix}\dot{e}\\ \dot{p}\end{pmatrix} &= \begin{pmatrix}0 & \frac{1}{\alpha}\\ \theta\beta & -\theta\left(\beta + \frac{\sigma}{\alpha}\right)\end{pmatrix}\begin{pmatrix}e - \overline{e}\\ p - \overline{p}\end{pmatrix}\\ &= \begin{pmatrix}0 & \frac{1}{\alpha}\\ \theta\beta & -\theta\left(\beta + \frac{\sigma}{\alpha}\right)\end{pmatrix}\begin{pmatrix}e\\ p\end{pmatrix} + \begin{pmatrix}-\frac{\overline{p}}{\alpha}\\ \theta\left(\beta + \frac{\sigma}{\alpha}\right)\overline{p} - \theta\beta\overline{e}\end{pmatrix}\end{aligned} \tag{8.21}$$

from which we clearly see that the equilibrium is defined when $\dot{e} = \dot{p} = 0$ are satisfied simultaneously and the equilibrium point is when $e = \overline{e}, p = \overline{p}$. To determine the local stability properties of the equilibrium point $(\overline{e}, \overline{p})$ of the system (8.21), we can follow the method in Chapter 5 and calculate the eigenvalues of the corresponding Jacobian J, which we can see directly from (8.21),

$$J = \begin{pmatrix}0 & \frac{1}{\alpha}\\ \theta\beta & -\theta\left(\beta + \frac{\sigma}{\alpha}\right)\end{pmatrix}. \tag{8.22}$$

The characteristic equation corresponding to the Jacobian (8.22) yields two eigenvalues that are given by

$$\mu_1 = \frac{1}{2}\left(-\theta\left(\beta + \frac{\sigma}{\alpha}\right) + \sqrt{\theta^2\left(\beta + \frac{\sigma}{\alpha}\right)^2 + 4\frac{\theta\beta}{\alpha}}\right),$$

$$\mu_2 = \frac{1}{2}\left(-\theta\left(\beta + \frac{\sigma}{\alpha}\right) - \sqrt{\theta^2\left(\beta + \frac{\sigma}{\alpha}\right)^2 + 4\frac{\theta\beta}{\alpha}}\right).$$

Since the parameters α, β, σ, $\theta > 0$ the two eigenvalues are real, one is negative and the other one positive. From the theory of dynamical systems we know that an equilibrium point $(\overline{e}, \overline{p})$ where the corresponding eigenvalues have such properties is a *saddle point*. In linear systems, a saddle point is characterised by its *manifolds*, or simply directional lines, in phase space that cross at the equilibrium point and determine the dynamics in the neighbourhood of the equilibrium point. The alignment of the manifolds is given by the eigenvectors of the Jacobian J and the *sign* of the corresponding eigenvalues describes the direction of the dynamics. A negative eigenvalue, in this context, is sometimes referred to as the *stable eigenvalue* since the direction of the dynamics on the corresponding manifold is pointing towards the equilibrium point. Similarly, a positive eigenvalue is sometimes referred to as the *unstable eigenvalue* since the direction of the dynamics on the corresponding manifold is pointing away from the equilibrium point.

The dynamics of the model and the time histories of the exchange rate and price are shown in Figure 8.1. Figure 8.1(a), which visualises the dynamics about the saddle point equilibrium, shows the two manifolds, the dashed lines cutting across the equilibrium point $(\overline{e}, \overline{p})$, and the direction of the dynamics on the manifolds are shown by the arrows. To understand the dynamics about the saddle point equilibrium, let us consider three initial conditions I, II and III, as seen in Figure 8.1(a). As can be seen from the directional arrows, among the three initial conditions only I, which is on the stable manifold, leads to a trajectory that ends up at the equilibrium point $(\overline{e}, \overline{p})$. The other two initial conditions, II and III, which do not lie on the stable manifold, lead to trajectories that initially follow the manifold and get closer to the equilibrium point but eventually diverge away from it. The dynamics highlights how a small change in the initial condition can have a large impact on the future outcome – here with a small deviation in the initial condition, say, from I to either II or III, the system diverges away from the equilibrium, as shown by the arrows, and will lead to either a continuous fall or rise in prices and the exchange rate, respectively.

The economic argument and insight of 'overshooting' that Dornbusch articulated as an explanation for the volatility in the exchange rate market can be explained using Figure 8.1(a). Consider that the system is at an arbitrary position such as (e_0, p_0) in Figure 8.1(a).

Suppose that there is an unanticipated permanent increase in the money supply. The rational agent would expect the price level and exchange rate to increase and that the economy would move to the equilibrium position $(\overline{e}_0, \overline{p}_0)$. However, the economy cannot instantaneously move to the equilibrium position since prices are assumed to be 'sticky', i.e. they only adjust gradually. The question that then remains is how does the economy get to the equilibrium level, i.e. what is the dynamics of adjustment towards the equilibrium?

The central analytical tool in Dornbusch's model that takes the economy to an equilibrium is the exchange rate, which is assumed to be the jump variable. Let us see how the dynamics works in Figure 8.1(a). Since prices are assumed to be rigid, for any increase in nominal money supply the equilibrium in the money market requires an increase in the demand for real money balances, which can happen only if the interest rate on domestic currency bonds falls. From the UIP condition (8.4) we can see that the domestic interest rate differential vis-a-vis the world interest rate ($r - r^*$), due to the fall in the home interest rate, will be offset by the expected appreciation of the home currency. The question is how does this happen since rational agents know that money is neutral and an increase in the money supply must lead to a proportionate depreciation in the exchange rate. Dornbusch's fundamental insight is that the initial depreciation (jump) of the exchange rate must be larger than the equilibrium level to leave room for the ensuing appreciation required to clear the bond and money markets, i.e. the exchange rate must overshoot the equilibrium level and then gradually adjust back to the equilibrium $(\overline{e}, \overline{p})$ during the period of the interest rate differential. Thus, starting from the initial equilibrium position (e_0, p_0) in Figure 8.1(a), the exchange rate depreciates from $(\overline{e}_0)$ by 'jumping' to point I, which is on the stable manifold, and then gradually appreciates back to the new equilibrium point $(\overline{e}, \overline{p})$. Note that point I is on the same price level as p_0 and only the exchange rate e has moved to a level higher than the equilibrium level of $\overline{e}$.

The idea of overshooting hinges on the assumption that exchange rates can jump easily, within seconds, of the arrival (or even in the anticipation) of 'news'. Also, it crucially depends on the stickiness of prices (and wages) that do not change immediately, as fast as the exchange rate. However, the dynamics of the model is such that the exchange rate has to jump exactly onto the stable manifold, i.e. to the point I, to take the system to the equilibrium. A small slip-up in the jump, so that the exchange rate lands on either II or III, will lead to dynamics that moves away from the equilibrium. The time histories of $e(t)$ for the three trajectories with initial conditions at I, II and III are shown in Figure 8.1(b). As can be seen from the figure, only the initial condition I, on the stable manifold, allows the system to reach the equilibrium $(\overline{e}, \overline{p})$. The neighbouring points II and III lead the system away from the equilibrium. Therefore, the discrete jump in exchange rate e has to be very precise for it to land exactly on the stable manifold in order for the system to converge to the equilibrium. While economic intuition makes sense, the arbitrary nature of the out of equilibrium adjustment, i.e. a discrete jump in a continuous model, makes the model intractable from a mathematical point of view.

To simulate and visualise the Dornbusch model we can follow the method used for the IS-LM model in Chapter 5. The MATLAB® code

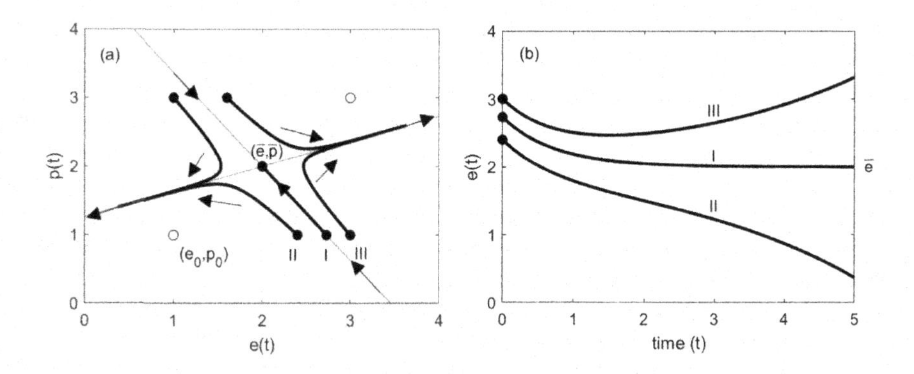

Figure 8.1: Trajectories of the Dornbusch system for α = 1, β = 1, σ = 1 and θ = 0.5 with the saddle point located at $(\overline{e}, \overline{p}) = (2,2)$ are plotted in (a), where the initial conditions are I: (2.7321,1), II: (2.4,1), III: (3,1). In (b), time histories for $e(t)$ are plotted for initial conditions I, II and III in (a). The thin lines in (a) indicate the manifold location and the arrows show direction of the dynamics in forward time.

to simulate the system of the two first-order differential equations (8.21) is given below in five separate parts, namely, (i) the initiation of the simulation and parameter values, (ii) the initial conditions, the call of the differential equation solver `ode45` and the output, (iii) the Dornbusch system, (iv) the calculation of the eigenvalues and eigenvectors, and (v) the plotting of the trajectories ($e(t)$, $p(t)$) and time histories of $e(t)$ and $p(t)$, as seen in Figure 8.1(b). The first set of code clears the memory, closes all the figures, sets the parameters and equilibrium point and creates the Jacobian J.

Initiation and parameter values

```
%Dornbusch overshooting model
clear all
close all
format compact

%Parameters
alpha  = 1;
beta   = 1;
sigma  = 1;
theta  = 0.5;

J = [0 1/alpha ;theta*beta -theta*(beta+sigma/alpha)];
```

```
%Equilibrium
e_eq = 2;
p_eq = 2;
```

Next we set up the simulation and decide simulation time, initial conditions, solver properties. The output from the ODE solver is a time vector t and 2-column state matrix Y, where the first column is e and the second column is p. Finally, we separate the output in an E- and P-vector for the plotting.

Initial condition and the ODE solver

```
%Simulation time
t_start = 0;
t_end = 5;

%Initial conditions
e0 = 3;
p0 = 1;
%Simulation tolerances
options = odeset('RelTol',1e-6,'AbsTol',1e-6);
[t,Y] = ode45(@(t,y) dornbusch_system(t,y,p),...
          [t_start t_end],[e0;p0],options);
%Output
E = Y(:,1);
P = Y(:,2);
```

The ODE solver calls the function `dornbusch_system(t,y,p)` which evaluates the right-hand side of (8.21), with time (t), state variables (y), equilibrium values (`e_eq` and `p_eq`) and the Jacobian (J) as input.

The Dornbusch system

```
function dedp = dornbusch_system(t,y,e_eq,p_eq,J)

    %Variables
    e = y(1);
    P = y(2);
```

```
    %The ODE model
    dedp = J*[e-e_eq; p-p_eq];
end
```

In order to calculate the eigenvalues and eigenvectors of the Jacobian J, which were discussed but not shown above, we can use the following MATLAB® code.

Eigenvalues and eigenvectors

```
%Eigenvalues
[eigenvectors,eigenvalues] = eig(J);
mu_1 = eigenvalues(1,1)
mu_2 = eigenvalues(2,2)

ev_1 = eigenvectors(:,1);
ev_2 = eigenvectors(:,2);
```

To visualise the output of the simulation, we follow the same methodology as in earlier chapters. The new thing here is that we use the eigenvectors `ev_1` and `ev_2` for plotting the stable and unstable manifolds.

Visualisation

```
%Plotting
figure(1)
subplot(2,1,1), hold on
plot(t,E,'k','LineWidth',2)
plot(0,e0,'o','MarkerFaceColor','k',...
    'MarkerEdgeColor','k')
xlabel('time (t)'),ylabel('e(t)')
hold off, box on
subplot(2,1,2), hold on
plot(t,P,'k','LineWidth',2)
plot(0,p0,'o','MarkerFaceColor','k',...
```

```
    'MarkerEdgeColor','k')
xlabel('time (t)'),ylabel('p(t)')
hold off, box on

figure(2), hold on
plot(E,P,'k','LineWidth',2)
xlabel('e(t)'),ylabel('p(t)')
k1 = ev_1(2)/ev_1(1);
k2 = ev_2(2)/ev_2(1);
plot([0 4],k1*([0 4]-2)+2,'k-')
plot([0 4],k2*([0 4]-2)+2,'k--')
plot(2,2,'o','MarkerFaceColor','k',...
    'MarkerEdgeColor','k')
plot(e0,p0,'o','MarkerFaceColor','k',...
    'MarkerEdgeColor','k')
axis([0 4 0 4])
hold off, box on
```

8.4 CONCLUDING REMARKS, EXTENSIONS AND CHALLENGES

In this chapter, we discussed the role of expectations in exchange-rate markets. Rudiger Dornbusch developed the theory of exchange rate dynamics in 1976 with an aim to explain the volatility in the exchange-rate market particularly in the context when the advanced economies moved from a fixed exchange rate to a fully flexible exchange rate regime. Dornbusch introduced the idea of so called 'overshooting', which centred on the fact that the exchange rate is a variable whose timescale is much faster than other variables in the economy and reacts, or jumps, within seconds of news arriving, or sometimes in anticipation of the news. In fact, this is also true of the prices of other financial assets like securities, shares and bonds, and also goods traded in the commodity futures exchange market such as oil and wheat. By contrast, the prices of most goods and services, such as labour or capital (rent), do not adjust as quickly (or jump) like financial assets. Dornbusch's idea of overshooting rests on the interaction between prices with different timescales – one set of prices jump quickly in relation to 'news' and others do not. More fundamentally, the model postulates that overshooting in the exchange-rate market compensates for the sticky or sluggish prices of goods and services in the product market, and hence the former markets are more volatile than the latter.

The idea of overshooting is also integral to monetary policy, particularly the inflation targeting policy. Monetary policy makers take into account the impact of policy announcements on movements, or volatility, in the exchange-rate market. However, such a policy analysis depends on a number of underlying assumptions such as the economy in question is well integrated into the international financial market, i.e. an open economy with a flexible exchange rate, the agents are rational, particularly in the foreign exchange market, the economy in question is a price taker in the sense it cannot unilaterally influence the world interest rate, and on the sluggish adjustment of wages and prices. While overshooting may be seen as a short-term phenomenon in the exchange-rate market, there are instances in the history that it has had real long-term economic consequences. One such example is Thatcher's 1979 anti-inflationary monetary policy in the UK of reducing the growth of the money supply, which was expected to produce an equivalent amount of reduction in inflation. The prediction of the quantity theory of money-based model did not turn out to be true during that period. Instead, in anticipation of high interest rates to reduce inflation, the British pound appreciated sharply following the tight money policy. This led to an increase in unemployment, peaking at around 12 percent in 1983–1984, with the consequence of an appreciated real exchange rate and a higher real interest rate, and it took a few years for the economy to return to its trend level. This episode reminds us of the influence monetary variables such as exchange rates have on the real production economy, and the need to develop dynamic models of interaction between the two spheres, particularly in the current context of globally integrated financialised economies.

NOTES

1 Note that here we adapt Gandolfo's characterisation of the Dornbusch model. For a full elaboration of the model, see [2].

2 Note that we are dealing with a private economy without a government.

BIBLIOGRAPHY

[1] Dornbusch, R. (1976), Expectations and Exchange Rate Dynamics, *Journal of Political Economy,* Vol. 84, pp. 1161–1176.

[2] Gandolfo, G. (1997). *Economic Dynamics,* Springer Science & Business Media.

9 THE SOLOW-SWAN GROWTH MODEL

9.1 ECONOMIC PROBLEM

In the preceding chapters while studying Keynesian macroeconomic dynamics (Chapters 4 and 5), we looked at the economy in the short-run where the capital stock was held constant even when the level of investment increased (or decreased). This was indeed one of the inconsistencies in Keynes's *General theory* [5] as pointed out by Pigou [6]. Any change in aggregate investment has two effects on the economy – one on the demand side and other on the supply side. The demand side impact is well known as the *multiplier effect*, where changes in investment lead to a change in output through a multiplier. However, the change in investment also leads to a change in the productive capacity, or change in the capital stock, which changes the potential supply capacity of the economy.

Ever since the publication of the *General theory*, there were early attempts to extend the model in a long-run setting. For instance, Roy Harrod [3], a contemporary of Keynes in Cambridge University (UK), and Evsey Domar [2] in the USA, who, independently and almost concurrently, advanced such models of growth, which came to be known as the *Harrod-Domar model* in the Keynesian literature. The Harrod-Domar model provided the initial conditions for the emergence of the modern growth theory in macroeconomics. We will briefly outline the Harrod-Domar model before we study the Solow growth model, which forms the foundation for the modern growth theory. We shall simply refer to the Harrod-Domar model as the *Harrod model* for presentational simplicity.[1]

Harrod's model is a dynamical system that describes equilibrium growth (or a *warranted growth rate*), consistent with the ex-post saving-investment equality. In other words, the dynamical system generates a unique rate of growth that is compatible with firms' optimal rate of utilisation of capital. However, Harrod showed that this equilibrium is unstable, i.e. a deviation from the equilibrium will make the economy either explode or implode, and does not guarantee full employment.

DOI: 10.4324/9780429324406-9

To see the key insight of Harrod's model, let us assume a closed economy without government and abstracting from other components of aggregate demand except private investment. The level of output is driven by aggregate demand and the goods market equilibrium is given by the multiplier relation

$$Y = \frac{1}{s} I, \quad 0 < s < 1, \tag{9.1}$$

where Y represents total output, I is investment, s denotes the saving rate, or marginal propensity to save, and the multiplier is here given by $\frac{1}{s}$.

However, investment decisions are driven by demand dynamics (or in other words, expected profitability from employing additional capital stock) consistent with the accelerator principle. Since investment increases the productive capacity of the economy, the latter should have supply side implications for the goods market equilibrium. Taking the increase in the productive capacity into consideration, the system achieves a 'steady' growth when the aggregate demand grows at the same rate as the increase in the productive capacity of the economy. This can be seen by rewriting (9.1) as

$$\frac{I}{Y} = s \tag{9.2}$$

and decomposing the investment-output ratio as

$$s = \frac{I}{K} \frac{K}{Y}, \tag{9.3}$$

where $\frac{I}{K}$ is the rate of capital accumulation and $\frac{K}{Y}$ is the capital-output ratio. The latter ratio is assumed to be a constant v, so that $v = \frac{K}{Y}$ (where the coefficient v describes a fixed production function), and the rate of capital accumulation, which is the rate of capacity growth, is denoted by g. At equilibrium, the rate of capital accumulation, or the capacity growth, will equal the rate of growth in output, i.e.

$$\frac{I}{K} \equiv \frac{\Delta K}{K} = \frac{\Delta Y}{Y} = g. \tag{9.4}$$

Using (9.4), we can rewrite the investment output and saving relation as

$$\frac{I}{K} \frac{K}{Y} = s \quad \text{or} \quad gv = s. \tag{9.5}$$

Therefore, the condition for the steady-state growth in Harrod's model is

$$g = \frac{s}{v}. \tag{9.6}$$

Harrod and Domar originally held both s and v constants, since they are determined by institutional structures in the economy, and called the ratio $g_w = \frac{s}{v}$ as the *warranted growth.* We can now see the economic intuition that underlies the so called 'knife-edge' dynamics of the model.

If the actual growth is slower than the warranted rate (i.e. when $g < g_w$), then excess capacity would be generated, i.e. the growth of an economy's productive capacity is outstripping growth in aggregate demand, and will induce firms to invest less, which in turn will reduce demand growth further leading to more excess capacity being generated.

On the other hand, if the actual growth is greater than the warranted rate (i.e. when $g > g_w$), then the aggregate demand growth outstrips the capacity growth, which will induce firms to increase capacity through new investment and in turn will increase demand, and thus further reinforce the excess demand.

This is the famous 'knife-edge' property of the Harrodian equilibrium. The growth path is stable only where the actual and warranted growth rates are equal, which is a path on which it is like being on a knife edge and any deviation from the path will make the system unstable where the system can either explode $\left(\text{when } g > \frac{s}{v}\right)$ or implode $\left(\text{when } g < \frac{s}{v}\right)$.

An interesting implication of the Harrod model is that there is no reason to believe that the dynamic equilibrium, i.e. when $g = g_w$, would guarantee full employment. To see this, let us assume that the labour force grows at some rate g_n, which Harrod called the *natural rate of growth,* since it refers to the maximum growth permissible by the population growth. Now, at any level of production technology, employment growth is proportional to output growth. Therefore, given the labour force growth g_n, the necessary condition for full employment in this economy is that the output must grow at the same rate as the labour force growth. In fact, this is the necessary condition both for full employment and for the stability of the employment growth. If the two rates are not equal, i.e. the rate of growth of output g and the rate of growth of labour force g_n, then the economy will eventually either run out of labour (when $g > g_n$) or battle with an ever-increasing unemployment rate (if $g < g_n$).

However, the only way these three growth rates, i.e., the actual growth g, the warranted growth g_w, and the labour force growth g_n, can coincide is by pure chance because the condition for dynamic equilibrium, $g = g_w$ is independent of the condition for a stable employment growth rate $g = g_n$. Further, since the parameters defining the warranted growth g_w, the saving rate s, the capital-output ratio v, and the labour force growth g_n are all exogenous and driven by different mechanisms, it is highly unlikely that the three rates would coincide, i.e. $g = g_w = g_n$, yielding to both a desirable and highly unlikely state of the economy referred to as the 'golden age' [7].[2]

Therefore, the question that remained unanswered in the context of Harrod's model was whether growth in the capitalist economies is precariously balanced on a knife-edge. There were several attempts to answer this question by both the followers of Keynes, the post-Keynesian economists, and the neoclassical economists. In the following, we will look at how the latter addressed the question using the famous Solow-Swan growth model.

9.2 MODELLING

As discussed in the previous section, in Harrod's model economic growth is unstable, both in terms of the dynamic equilibrium between actual growth and the warranted growth (where $g = g_w$) and the full-employment equilibrium, where the actual growth equals the growth in the labour force (and thus $g = g_n$). Robert M. Solow and Trevor Swan argued that the instability in Harrod's model arises due to the exogenously fixed capital-output ratio and proposed a growth model where the very same ratio becomes the adjusting variable that leads the system back to its growth path. This means that the parameter v would endogenously change to bring the system to the steady-state growth path. In other words, while dynamic equilibrium is assumed, with investment passively adjusting to available saving, the full-employment equilibrium is achieved through the infinitesimal substitutability between capital and labour, using a twice continuously differentiable production function that exhibits constant returns to scale. The model is often referred to as the *neoclassical* model because of the following reasons: (a) savings determines investment and (b) disequilibrium adjustment is explained by fully flexible and competitive factor markets. Let us now study the model in more detail.

The assumptions of the Solow-Swan growth model are as follows. The production function F with arguments of both capital K and labour L is given by

$$Y = F(K, AL), \tag{9.7}$$

where A represents technology/knowledge and AL is referred to as *effective labour*. The technology term written in this way reflects the *labour-augmenting*, or *Harrod-neutral*, nature of technical progress.[3] The production function F is assumed to exhibit constant returns to scale, or homogeneous, in both capital K and labour L. This means that doubling the quantities of inputs (with A fixed) doubles the output Y. Since we are interested in how output per labour grows over time (or in general, how output per capita grows over time), we can rewrite the production function in the *intensive form*, since the function is homogeneous as[4]

$$Y = ALF\left(\frac{K}{AL},1\right) \quad \Leftrightarrow \quad \frac{Y}{AL} = F(r,1), \quad r = \frac{K}{AL}, \tag{9.8}$$

where $\frac{Y}{AL}$ is the amount of output per unit of effective labour and r is the amount of capital per unit of effective labour. Letting $y = \frac{Y}{AL}$, (9.8) can be rewritten as

$$y = f(r) = F(r,1). \tag{9.9}$$

The production function $f(r)$ is assumed to satisfy

$$f(0) = 0, \quad \frac{df}{dr}(r) > 0 \quad \text{and} \quad \frac{d^2 f}{dr^2}(r) < 0$$

implying that the marginal product of capital is positive, but decreasing as capital increases. In addition there are conditions on the curvature of $f(r)$, namely,

$$\lim_{r\to 0^+} \frac{df}{dr}(r) = \infty \quad \text{and} \quad \lim_{r\to\infty} \frac{df}{dr}(r) = 0,$$

which are referred to as the *Inada conditions*. These conditions ensure that growth of the economy does not diverge when capital stock becomes large (or tends towards infinity).

Aggregate demand in this one-good, closed, and private economy is given by

$$Y = C + I, \tag{9.10}$$

where C denotes consumption and I denotes investment expenditures. The macroeconomic equilibrium condition for this economy is

$$S = I. \tag{9.11}$$

Further assuming that a part of income is being consumed and the rest is saved, the savings function can be written as

$$S = sY, \quad 0 < s < 1, \tag{9.12}$$

where s is the saving propensity.

The gross investment of the economy is defined as the change in capital stock plus the depreciation of the existing capital stock and is given by

$$I = \frac{dK}{dt} + \delta K, \quad 0 < \delta < 1 \tag{9.13}$$

where δ denotes the depreciation rate of capital stock. Finally, assume that both the labour supply L in the economy and the knowledge accumulation A grow at constant exponential rates n and a respectively as

$$L(t) = L_0 e^{nt}, \quad A(t) = A_0 e^{at}, \tag{9.14}$$

for some positive intial values L_0 and A_0.

Substituting (9.12) and (9.13) in (9.11) yields,

$$sY = \frac{dK}{dt} + \delta K. \tag{9.15}$$

Now we need to find the expressions for the left hand and right hand side of (9.15) to complete the model.

From (9.8) and (9.9) we get

$$Y = ALf(r), \tag{9.16}$$

which can be substituted in the left-hand side of (9.15).

In order to determine $\frac{dK}{dt}$ on the right-hand side, we use $r = \frac{K}{AL}$ (from (9.8)), which can be written as $K = ALr$ and upon total differentiation yields,

$$\frac{dK}{dt} = \frac{dA}{dt}Lr + A\frac{dL}{dt}r + AL\frac{dr}{dt} = aALr + nALr + AL\frac{dr}{dt}. \tag{9.17}$$

Finally, substituting (9.16) and (9.17) in (9.15) yields

$$sALf(r) = aALr + nALr + AL\frac{dr}{dt} + \delta ALr \tag{9.18}$$

and further simplification gives rise to the equation

$$\frac{dr}{dt} = sf(r,1) - (n + a + \delta)r, \tag{9.19}$$

which is the fundamental dynamic equation of Solow's growth model.

To further understand the dynamics and the stability properties of the model, we will use the so called Cobb-Douglas production as a special case for the production function $f(\mathrm{r})$, which is given by

$$Y = K^{\alpha}(AL)^{1-\alpha}, \quad 0 < \alpha < 1, \tag{9.20}$$

where α is the proportion of capital K, and $1-\alpha$ is the proportion of effective labour AL, used to produce the output. Expressing the Cobb-Douglas production function (9.20) in terms of output per effective labour, i.e. dividing by AL, yields

$$\frac{Y}{AL} = \left(\frac{K}{AL}\right)^{\alpha} \Rightarrow y = r^{\alpha} \tag{9.21}$$

and the production function for this specific case becomes $f(r) = r^{\alpha}$.

Therefore, using (9.21) in (9.19) yields

$$\frac{dr}{dt} = sr^{\alpha} - (n + a + \delta)r, \tag{9.22}$$

which is the fundamental dynamic equation of Solow's growth model for the special case of Cobb-Douglas production function. Note that equation (9.22) is a first-order nonlinear differential equation, whose dynamic properties will be explored in the following section.

9.3 ANALYSIS, SIMULATION AND VISUALISATION

As mentioned in the previous section, the fundamental equation in Solow's model (9.22) is a first-order nonlinear differential equation, which usually means that no analytical solution is available for further analysis. However, this is one of those cases where it is possible to transform the nonlinear system to a linear one through a variable transformation and the linear system can be solved analytically. In what follows we will first show how this transformation can be done and what the solution looks like. Second, we will use traditional linear stability analysis methods for (9.22) to find and analyse the stability of all equilibrium points. Finally, we will show a graphical method to locate and determine the stability of the equilibrium points. After these three methods we will show how to numerically solve (9.22) using MATLAB®.

STABILITY ANALYSIS THROUGH BERNOULLI TRANSFORMATION

Equation (9.22) can be made linear by a simple transformation, known as the *Bernoulli transformation*, by defining a new variable

$$\kappa = r^{1-\alpha}. \tag{9.23}$$

Note that this mathematical transformation has an economic interpretation, since from (9.20) we have

$$\frac{K}{Y} = \left(\frac{K}{AL}\right)^{1-\alpha} = r^{1-\alpha} = \kappa, \tag{9.24}$$

where κ transforms the non linear system (9.22) in capital/effective labour (K/AL) ratio to a linear system in capital/output (K/Y) ratio owing to the use of the Cobb-Douglas production function (9.20).

We can now restate the fundamental dynamic equation of Solow's model (9.22) in lieu with the transformation variable κ, as follows. First, we differentiate $\kappa = r^{1-\alpha}$ with respect to time and get

$$\frac{d\kappa}{dt} = (1-\alpha) r^{-\alpha} \frac{dr}{dt} \quad \Rightarrow \quad \frac{1}{1-\alpha}\frac{d\kappa}{dt} = r^{-\alpha}\frac{dr}{dt}. \tag{9.25}$$

Now multiplying both sides of (9.22) by $r^{-\alpha}$ yields,

$$r^{-\alpha}\frac{dr}{dt} = s - (n + a + \delta) r^{1-\alpha} \tag{9.26}$$

and using equations (9.25) and (9.26) we have

$$\frac{1}{1-\alpha}\frac{d\kappa}{dt} = s - (n + a + \delta)\kappa \tag{9.27}$$

or

$$\frac{d\kappa}{dt} + (n + a + \delta)(1-\alpha)\kappa = s(1-\alpha), \tag{9.28}$$

which is a linear differential equation in κ with constant coefficients. Assuming that $\kappa = \kappa_0$ at $t = 0$, the general solution to (9.28) is

$$\kappa(t) = \left(\kappa_0 - \frac{s}{n+a+\delta}\right)e^{-(n+a+\delta)(1-\alpha)t} + \frac{s}{n+a+\delta}. \tag{9.29}$$

Since $n + a + \delta$ and $1 - \alpha$ are positive, the term $e^{-(n+a+\delta)(1-\alpha)t} \to 0$ as $t \to \infty$ and $\kappa(t)$ tends to its equilibrium value

$$\kappa^* = \frac{s}{n+a+\delta}.$$

Furthermore, recall that the difference $(\kappa_0 - \kappa^*)$ tends to 0 at a rate given by

$$\beta = (n+a+\delta)(1-\alpha), \tag{9.30}$$

where β is sometimes referred to as the *coefficient of convergence*.

Finally, the solution for the original system (9.22), using the transformed equation (9.28), is given by

$$r(t) = \kappa^{\frac{1}{1-\alpha}} = \left(\left(r_0 - \frac{s}{n+a+\delta}\right)e^{-(n+a+\delta)(1-\alpha)t} + \frac{s}{n+a+\delta}\right)^{\frac{1}{1-\alpha}}, \tag{9.31}$$

where we used $\kappa_0 = r_0^{1-\alpha}$ and the equilibrium, or the steady-state value of capital to effective labour ratio, r^* is

$$r^* = \left(\frac{s}{n+a+\delta}\right)^{\frac{1}{1-\alpha}}. \tag{9.32}$$

LINEAR STABILITY ANALYSIS

Instead of using the Bernoulli transformation we can analyse the original system (9.22) directly by first locating all equilibrium points and analysing their stability.

First locate the equilibrium points r^* as

$$\frac{dr}{dt} = 0 \quad \Rightarrow \quad s\left(r^*\right)^{\alpha} - (n+a+\delta)r^* = sr^*\left(\left(r^*\right)^{\alpha-1} - \frac{n+a+\delta}{s}\right) = 0, \tag{9.33}$$

which yields two equilibrium points that are given by

$$r_1^* = 0 \quad \text{and} \quad r_2^* = \left(\frac{s}{n+a+\delta}\right)^{\frac{1}{1-\alpha}}.$$

The former equilibrium point is the trivial one, which is not picked up by the Bernoulli transformation above, and the latter is the same as the one given by the Bernoulli transformation. To calculate the stability of the equilibrium points we can differentiate the right-hand side of (9.22) with respect to r and evaluate at the two equilibrium points. If the sign of the differential is negative (positive), it implies that the equilibrium point is stable (unstable). Here we have,

$$\frac{d}{dr}\left(\frac{dr}{dt}\right) = \alpha s r^{\alpha-1} - (n+a+\delta), \tag{9.34}$$

and evaluating at the equilibrium points respectively yields,

$$\left.\frac{d}{dr}\left(\frac{dr}{dt}\right)\right|_{r=0} = \lim_{r\to 0^+} \frac{\alpha s}{r^{1-\alpha}} - (n+a+\delta) = \infty > 0, \tag{9.35}$$

$$\begin{aligned} \left.\frac{d}{dr}\left(\frac{dr}{dt}\right)\right|_{r=\left(\frac{s}{n+a+\delta}\right)^{\frac{1}{1-\alpha}}} &= \alpha s\left(\left(\frac{s}{n+a+\delta}\right)^{\frac{1}{1-\alpha}}\right)^{\alpha-1} - (n+a+\delta) \\ &= (\alpha-1)(n+a+\delta) < 0 \quad \text{for } 0<\alpha<1. \end{aligned} \tag{9.36}$$

The above expressions (9.35) and (9.36) show that $r_1^* = 0$ is unstable and $r_2^* = \left(\frac{s}{n+a+\delta}\right)^{\frac{1}{1-\alpha}}$ is stable.

In economic terms the stable equilibrium, where r (the ratio of capital to effective labour) reaches the equilibrium value r_2^*, is often referred to as the *balanced growth path*, the reason being that as the equilibrium value of capital stock K is equal to ALr_2^* (see (9.24)) and will grow at a constant rate given by $n + a$ (derived from (9.8)), which is the rate of growth of effective labour. With both capital and effective labour growing at the constant rate of $n + a$, output Y will also grow at the same rate. Consequently, at the stable equilibrium r_2^*, capital per effective worker $\left(\frac{K}{AL}\right)$, and output per effective worker $\left(\frac{Y}{AL}\right)$ will grow at the constant rate of $n + a$. This constant rate of growth is called the balanced or the steady state growth path.

GRAPHIC STABILITY ANALYSIS

Next we will use a graphical method to perform the analysis in the previous section. The graphical method is a popular presentation of the Solow model in undergraduate textbooks. Accordingly, we shall analyse the two terms on the right-hand side of (9.22) graphically and qualitatively compare them in order to determine whether $\frac{dr}{dt}$ is growing or declining. Let us denote the two terms as

$$y_1(r) = sr^{\alpha} \quad \text{and} \quad y_2(r) = (n + a + \delta)r,$$

and plot them in Figure 9.1.

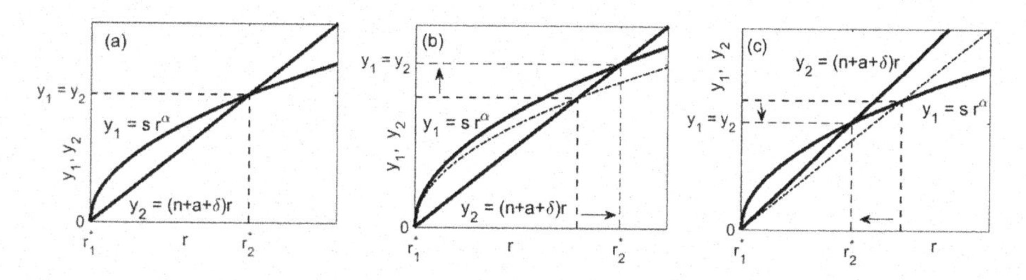

Figure 9.1 Schematics of graphical stability analysis, where (a) is the base case. In (b) s is increased that leads to an increase in r_2^* and in (c) n is increased that leads to a decrease in r_2^*.

We see that the functions are equal and cross at two separate points, r_1^* and r_2^*, where

$$\frac{dr}{dt} = y_1\left(r_k^*\right) - y_2\left(r_k^*\right) = 0, \quad k = 1, 2,$$

as expected, since it is the condition for an equilibrium point, as we saw in (9.33). Furthermore, in Figure 9.1 we see that for $r > 0$ we have the following two cases

$$\text{I:} \quad r_1^* < r < r_2^* \Rightarrow y_1 > y_2 \quad \text{and} \quad \frac{dr}{dt} > 0,$$

$$\text{II:} \quad r > r_2^* \Rightarrow y_1 < y_2 \quad \text{and} \quad \frac{dr}{dt} < 0.$$

The way to interpret these conditions is that $r(t)$ is increasing with time t if $r_1^* < r < r_2^*$ and decreasing if $r > r_2^*$, which means that r_1^* is repelling (unstable) and r_2^* is attracting (stable) equilibrium points respectively. That is, an economy with a low level of capital-labour ratio (r close to zero) will grow and an economy with a capital-labour ratio higher than the steady-state value (r_2^*) will contract. This is consistent with the above analytical methods, first through the Bernoulli transformation and (9.32) and second through the standard stability analysis (9.35) and (9.36). The MATLAB® code to generate Figure 9.1 is given in the following.

Graphical stability analysis

```
%Solow model
clear all
close all

%Parameters
s     = 0.8;
alpha = 0.5;
n     = 0.5;
a     = 0.3;
delta = 0.4;

%Variable r
r1 = linspace(0,1,100);
r2 = [0,1];

%Plotting
r_eq=(s/(n+a+delta))^(1/(1-alpha));

figure(10), hold on
plot(r1,s*r1.^alpha,'k','LineWidth',2)
plot(r2,(n+a+delta)*r2,'k','LineWidth',2)
plot(r_eq*[1,1],s*r_eq^(alpha)*[0,1],'k--')
plot(r_eq*[0,1],s*r_eq^(alpha)*[1,1],'k--')
xlabel('r'),ylabel('y_1,y_2')
axis([0,1,0,1])
box on, hold off
```

The graphical analysis shown in Figure 9.1 is often used in undergraduate textbooks to convey the underlying dynamics and other comparative static analysis of the model. Here we show the base equilibrium scenario (Figure 9.1 (a)) and two other scenarios where the saving rate s is increased and the other case where the depreciation of capital δ is increased (in Figures 9.1 (b) and (c), respectively). The shifts in the respective curves are shown with solid lines in these plots.

From a policy perspective, maximizing social welfare implies maximizing consumption per capita (or per worker). Therefore, policy makers would be more interested in the behaviour of consumption than in the output itself. So the obvious question is to find consumption per effective labour on the balanced growth path. In this model, consumption per unit of effective labour is obtained by multiplying the fraction of output $(1 - s)$ that is consumed with output per unit of effective labour $f(r)$. We know that an increase in s increases the steady-state capital per effective labour r^*. The question then is whether the increase in the savings rate raises or lowers the steady-state consumption. This would depend on whether the slope $f'(r)$ – the marginal product of capital – of the production function $f(r)$ is larger or smaller than $n + a + \delta$, the depreciation loci. In other words, if the additional capital (via higher savings) produces more output than the depreciation, then it would increase consumption. If the additional capital, via higher savings, produces less output than the depreciation loci, then consumption will have to decrease to maintain the steady-state growth. Therefore, the optimal level of consumption is when $f'(r)$ equals the depreciation loci and the steady-state capital stock (or capital-labour ratio). This scenario that generates the optimum level of consumption and is referred to as the *golden rule level of capital stock*.

SIMULATION AND VISUALISATION

In Chapters 5 and 6, we simulated and visualised first-order linear ordinary differential equations. We will follow the same methodology here for simulating the Solow model. Starting with the following MATLAB® code that sets up the scene.

Initiation and parameter values

```
clear all
close all
format compact

%Parameters
s     = 0.8;
alpha = 0.5;
n     = 0.3;
a     = 0.3;
delta = 0.4;
```

After initialising the parameters, we are ready to simulate the system by defining start and end times, an initial condition, simulation options and finally simulate the system using the MATLAB® function `ode45`.

Simulation

```
%Simulation time
t_start = 0;
t_end = 10;

%Initial conditions
r0 = 0.9;

% Simulation setup
options = odeset('RelTol',1e-6,'AbsTol',1e-6);

% Simulation
[t,R] = ode45(@(t,r) solow_system(t,r,s,alpha,...
        n,a,delta),[t_start t_end],r0,options);
```

The function ode45 calls the function `solow_system` that is given by the following MATLAB® code.

The ODE system

```
function dr = solow_system(t,r,s,alpha,n,a,delta)

%The ODE model
dr = s*r^alpha-(n+a+delta)*r;
```

Finally we plot the time history of $r(t)$ and also indicate where the stable equilibrium value lies using the following MATLAB® code.

Visualisation

```
%Plotting
r_eq = (s/(n+a+delta))^(1/(1-alpha));

figure(21), hold on
plot(t,R(:,1),'k','Linewidth',2)
plot([t_start,t_end],r_eq*[1,1],'k--','Linewidth',1)
axis([0,10,0,1]), box on
xlabel('t'), ylabel('r(t)')
```

In Figure 9.2 we plot the time history of $r(t)$, the ratio of capital to effective labour, for two different initial conditions – think of two economies with different starting points. Figure 9.2 (a) is the baseline scenario and the other plots, Figures 9.2 (b) and (c), show when α, the marginal product of capital per effective labour (the capital's share in national income) is increased and when the population growth rate n is increased respectively. Two observations are in order. First, the increase in capital's share α, or the marginal product of capital, increases the steady-state growth (Figures 9.2 (b)). Second, when the population growth rate is increased, the steady-state value declines (Figures 9.2 (c)), which is in line with what we saw in scenario in Figure 9.1 (c). Third, in all the scenarios we have seen the economy with a low capital to effective labour as a starting point ($r(0) = 0.01$) grows faster than the economy with a high capital to effective labour ($r(0) = 0.9$). This is one of the fundamental insights of Solow's growth model and influenced the question of 'convergence' and attracted considerable empirical work in the economic-growth literature, which investigates whether poor countries tend to grow faster than rich countries and how long it would take the former to catch up with the latter. One of the reasons for such convergence is the flow of capital from rich countries to the poor countries where the rate of return on capital would be greater because of low level of capital per worker. Also, technological development and diffusion of knowledge that comes along with it, albeit with a lag, will help poor countries to grow faster and help them converge with the richer countries. Solow's growth model opened up a huge area of empirical research in this area, which has since dominated the debate on growth differentials between countries around the world.

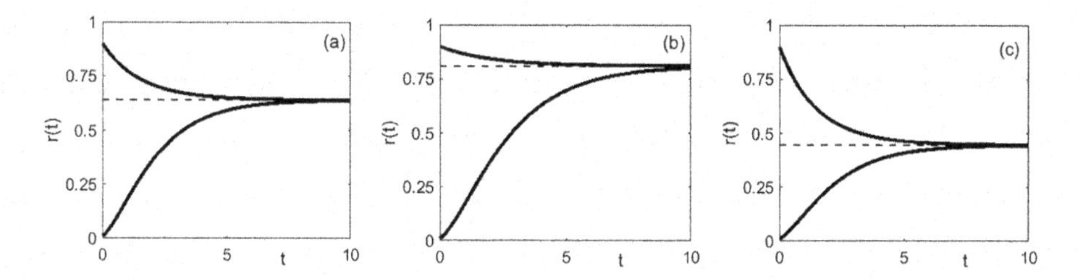

Figure 9.2: Time histories of the capital-labour ratio ($r(t)$) for different parameter values. In (a) $s = 0.8$, $\alpha = 0.5$, $n = 0.3$, $a = 0.3$ and $\delta = 0.4$, while in (b) $\alpha = 0.9$ and (c) $n = 0.5$. In all three cases two different initial conditions are used, namely $r(0) = 0.01$ and $r(0) = 0.9$.

9.4 CONCLUDING REMARKS, EXTENSIONS AND CHALLENGES

In this chapter we studied Solow's growth model, which is considered as the fundamental growth model in the modern growth literature. Robert Solow's aim was to provide a solution to Harrod's instability problem, or the knife-edge instability, by endogenising the capital-output ratio, which would bring the system to steady-state growth. In other words, Solow proposed a model where investment was passively adjusting to available saving and the full-employment equilibrium is achieved through the variable capital-labour ratio via the neoclassical production function mechanism. This is one of the reasons why Solow's model is often referred to as the *neoclassical growth model*.

The Solow model articulated the limit to the then dominant view that accumulation of physical capital can create sustained economic growth and also helped understanding of the cross-country income differences. Solow's model showed the inevitability of reaching a stationary state for such physical capital accumulation driven growth process and pointed to the central role of effectiveness of labour, or technology A, in generating permanent growth in output per worker. Solow further argued that the differences in the effectiveness of labour are fundamental for understanding the cross-country growth and wealth differences temporally and spatially. However, Solow's model assumed the technology term as a given constant and neither explored its determinants nor examined how it creates conditions for permanent growth. It remains a catch all or a residual term for factors other than labour and capital that affect economic growth – it is referred to in the literature as the *Solow residual*. That task was taken up by models referred to as *endogenous growth models*. In this family of models, the effectiveness of labour or technology corresponds to abstract knowledge. Intuitively, these models argue that economic growth is driven by the evolution of knowledge over time and the cross-country differences in real incomes could be explained by the availability, or lack, of education, skills of the labour force, etc. In the next chapter, we will study a general model in the family of endogenous growth models.

NOTES

1 See Daniele Besomi [1] for the history of Harrod's model.
2 The period 1945–1973, in which full employment actually did prevail in most advanced capitalist economies came to be known as the 'golden age of capitalism' (cf. Marglin and Schor 1990) [4].

3 If technology augments the capital input as in $Y = F(AK,L)$, then it is called *capital augmenting* or *Hicks-neutral* technical progress.
4 See Romer [8] for a more detailed description of the mathematical derivations.

BIBLIOGRAPHY

[1] Besomi, D. (1997), Roy Harrod and Traditional Theory, *European Journal for the History of Economic Thought,* Vol. 4, No. 1, pp. 92–115.

[2] Domar, E. (1946), Capital Expansion, Rate of Growth, and Employment, *Econometrica,* Vol. 14, No. 2, pp. 137–147.

[3] Harrod, R.F. (1939), An Essay in Dynamic Theory, *The Economic Journal,* Vol. 49, pp. 14–33.

[4] Marglin, S.A. and Schor, J.B. (Eds.) (1991), *The Golden Age of Capitalism: Reinterpreting the Postwar Experience,* Oxford University Press.

[5] Keynes, J.M. (1936), *The General Theory of Employment, Interest and Money,* London: Macmillan.

[6] Pigou, A.C. (1936), Mr. JM Keynes' General Theory of Employment, Interest and Money. *Economica,* Vol. 3, No. 10, pp. 115–132.

[7] Robinson, J.V. (1956), *The Accumulation of Capital,* Macmillan: London.

[8] Romer, D. (2012), *Advanced Macroeconomics (4th ed.),* New York: McGraw-Hill.

10 AN ENDOGENOUS GROWTH MODEL

10.1 ECONOMIC PROBLEM

In Chapter 9, we looked at the Solow-Swan growth model in some detail (henceforth the 'Solow model'). The central result of the Solow model is that saving drives growth, but this is temporary, as the economy tends to a stationary state in the long run, due to the diminishing returns to capital. In other words, increasing savings can only deliver 'scale' effects but not 'growth' effects, i.e. the scale of output growth increases in the short term but there is no increase in the long-run growth rate. Such a result runs counter to the Keynes-inspired growth models where capital accumulation – the independent role of investment – is the prime mover, similar to what we saw in the Harrod-Domar model and other Keynesian models. Solow's model shows the limit to the capital accumulation led growth process whereby the economy hits a stationary state in the long run.

Solow's model inspired a huge empirical literature on the so called *convergence hypothesis*. At the most basic level, the question was about how the poor countries can grow faster and catch up with the rich countries. From the perspective of Solow's model there are three reasons why one might expect such convergence. First, the model predicts that countries converge to their steady-state growth and depending on their relative difference in the initial conditions of capital per worker, one would expect poor countries to grow faster and catch up with the rich countries. Second, the Solow model implies that the rate of return on capital is lower in countries with more capital per worker and thus there are incentives for capital to flow from rich to poor countries, which tends to help convergence. Third, differences between growth rates may be due to the lack of availability of production technologies and these differences might tend to shrink as knowledge diffuses to the poor countries and improve their rate of growth.

However, the empirical literature, for instance see [4], fails to explain the growth differentials across countries purely in terms of the variations

DOI: 10.4324/9780429324406-10

in capital per effective worker $\left(\frac{K}{AL}\right)$. Neither the empirical differences in capital per worker across countries nor the relationship between output per worker and capital per worker suggest the role of capital per worker in explaining variations in countries' growth rates [5, 6]. Further, as noted above, growth differentials in output per worker on the basis of capital per worker implicitly imply significant differences in the rate of return on capital across countries over time and space. If this is true, we would observe huge flows of capital from rich to poor countries. However, empirical literature highlights evidence of capital flows in the opposite direction [3].

The question then is since capital accumulation (capital per worker) neither explains the permanent increase in the steady-state growth rate (the *growth effect*) of individual countries nor the cross-country income differentials, what other factors could explain these central questions of economic growth? In the Solow model, the only other factor that could come into play is the 'effectiveness of labour', i.e. the technology term A. However, the term remains exogenous to the model and represents all the residual factors other than labour and capital that could affect output. Notably, in the growth accounting exercises, pioneered by Abramovitz [1] and Solow [9], it is known as the *Solow residual*.

The next logical step is to understand whether the effective labour or technology can create any growth effect, and what causes its variations temporally and spatially. In other words, one needs to study how technology evolves endogenously with the economy to create the so called growth effect. In the economic growth literature, the class of models that endogenise the technology term have been collectively known as the *endogenous growth models,* where the focus is on *knowledge production* as the fundamental determinant of technology, and these models strive to understand the cross-country growth differentials in output per worker in terms of knowledge accumulation and diffusion.

Two seminal papers by the Nobel prize winning economist Paul Romer laid the foundations for the development of the endogenous growth theory. Romer's key insight is that knowledge production is fundamentally driven by 'ideas' (Romer [7], [8]). According to Romer [8], an idea is a design, or a blueprint, or a set of instructions for producing something new, or transformation of an existing object in some way that generates more output or utility. Examples may include a formula for a new vaccine, mathematical tools such as algebra and calculus, or the design of the QWERTY keyboard layout. Romer argued that the *nonrival* nature of ideas generates increasing returns to a scale that can then lead to sustained growth. Let us unpack the intuition behind Romer's argument.

The standard 'goods' used in economics, for instance in the utility or production functions, are *rivalrous,* i.e. as more people consume or use them there are fewer of these goods at any given level of supply. Examples of rival goods include apples, oranges, durable goods and services of a specialist neurosurgeon etc. The rivalrous nature of such goods and services is what underlies the *scarcity principle* in economics.

However, 'ideas' are nonrival. For example, consider the case of a computer operating system such as Linux, Windows and Unix, or the Internet and telecommunication signals, where even as more and more people use them, they do not immediately become scarce for others. Romer's favorite example was oral rehydration therapy, which was based on the idea of dissolving a few minerals such as salt and sugar in the right proportion to produce a solution that rehydrates children. Once this idea, or chemical formula, was discovered it can be used over and over again for years without becoming scarce.

How can the nonrivalrous nature of ideas explain economic growth? Think of the standard production function with constant returns to scale. If we want to double the output production, we need to double the inputs (labour and capital). However, once the design or blueprint for making a nonrival good is produced, firms do not need to reinvent the idea each time for building a new computer, aircraft, or a programming language. Instead firms can use the same idea and produce an increasing number of nonrival goods, be it advanced versions of the same product, and derive increasing returns to scale. Thus, taking both the constant return to scale of the rival goods and the increasing returns to scale of ideas in the standard production process into consideration, it is easy to see how growth would follow naturally. Since the stock of ideas, or knowledge, does not need to be divided among all people in the economy, unlike capital in the Solow model, increasing the stock of knowledge leads to an increase in output per person. We can see this argument analytically in the simple model referred to as the *AK model* that is expressed as

$$Y(t) = AK(t), \tag{10.1}$$

$$\dot{K}(t) = sY(t) - nK(t), \tag{10.2}$$

where Y is output, A is an exogenously given index of the level of technology (or the stock of knowledge), s is an exogenous saving (and investment) rate, n is the constant rate of population growth and K is the physical capital that can be interpreted as human capital as in Lucas (1988)[2]. Substituting the *knowledge production function* (10.1) into the capital accumulation equation (10.2) and defining g_K as the relative change in capital yields

$$gK = \frac{\dot{K}(t)}{K(t)} = sA - n. \tag{10.3}$$

The fundamental insight of the *AK* model, which is a first-order linear differential equation of the form

$$\frac{dK}{dt}(t) = (sA - n)K(t) = g_K K(t),$$

is that the increase in the stock of knowledge could generate exponential growth in the economy.[1]

Romer's insight inspired many extensions and modifications and laid the foundation for the development of the *endogenous theory* (also known as the *new growth theory*) as well as influenced growth policies around the world. The question is how the stock of knowledge varies over time as the flow of new ideas, i.e. $\frac{dA(t)}{dt}$, add to it and permanently increase the rate of per capita growth and produce the so called *growth effect.* We will answer this question in the next section by using Romer's model [8], as exposited in [6], where both capital accumulation and knowledge accumulation interact, and study how it generates permanent growth effect.

10.2 MODELLING

For the model we will develop, consider a closed economy without a government. There are two sectors in the economy, a goods-producing sector where output Y is produced and a research and development (R&D) sector where technology A (knowledge) is produced.[2] In terms of shares of labour and capital employed in these sectors, a fraction a_L $(0 \le a_L \le 1)$ of the labour force is used in the R&D sector and the remaining fraction $1 - a_L$ is used in the goods producing sector. Similarly, the R&D sector employs a fraction a_K $(0 \le a_K \le 1)$ of the total capital stock and the remaining fraction $1 - a_K$ is employed in the goods producing sectors. We assume that the shares a_L and a_K are exogenously given constants.

The production function in the goods-producing sector is now given by

$$Y(t) = \left[(1 - a_K)K(t)\right]^{\alpha}\left[A(t)(1 - a_L)L(t)\right]^{1-\alpha}, \quad 0 \le \alpha \le 1, \tag{10.4}$$

where L and K represent labour and capital, respectively, and the parameter α is a constant. The coefficients a_K and a_L represent the fraction of capital and labour employed, respectively, as defined above. Note that the goods production function written in the Cobb-Douglas form implies constant returns to both capital and labour.

Next, the production of new ideas depends on the quantities of capital and labour engaged in research and on the existing level of technology (knowledge), and is given by

$$\dot{A}(t) = B\left(a_K K(t)\right)^{\beta} \left(a_L L(t)\right)^{\gamma} A(t)^{\theta}, \quad B > 0, \beta \geq 0, \gamma \geq 0, \theta \in \mathbb{R} \tag{10.5}$$

where B, β and γ are exogenously specified constants. Note that the knowledge production function $A(t)$, given the parameter restrictions, is not assumed to obey constant returns to scale of capital and labour. The parameter θ captures the influence of existing stock of knowledge on the success of new knowledge, or the output of the R&D sector, and there is no restriction placed on θ as the influence of the existing stock of knowledge can be positive or negative for generating new knowledge. For instance, it is always harder to make new discoveries when the stock of knowledge is greater, and so θ can be both negative and positive.

Similar to what we had in Solow's model (Chapter 9), we assume the labour force to grow at a constant rate n so that

$$\dot{L}(t) = nL(t), \quad n > 0. \tag{10.6}$$

We also assume that the savings rate s is an exogenous constant, but do not allow depreciation, and thus the capital accumulation is given by

$$\dot{K}(t) = sY(t), \quad s > 0. \tag{10.7}$$

Using (10.4), (10.5) and (10.7), we can rewrite and define the goods production function g_K and the knowledge production function g_A as

$$g_K(t) \equiv \frac{\dot{K}}{K} = s\left(1 - a_K\right)^{\alpha} \left(1 - a_L\right)^{1-\alpha} \left(\frac{A(t)L(t)}{K(t)}\right)^{1-\alpha}, \tag{10.8}$$

$$g_A(t) \equiv \frac{\dot{A}}{A} = B\left(a_K K(t)\right)^{\beta} \left(a_L L(t)\right)^{\gamma} A(t)^{\theta-1}, \tag{10.9}$$

respectively. Note that the sign restrictions on $K(t) > 0$, $L(t) > 0$ and $A(t) > 0$, for all t, from (10.5) yield $g_A > 0$ and $g_K > 0$ in (10.8) and (10.9), respectively. This makes the analysis both economically meaningful and easier.

Note on mathematical analysis

It is worth pointing out that even if the mathematical analysis allows for a wide range of numerical values, it is pertinent to identify the set of values that is meaningful, i.e., economically meaningful in our case, in terms of the context for which it is intended. Notably, many models allow for a more general analysis than they are intended for.

Finally, we take natural logarithms of (10.8) and (10.9), differentiate with respect to time and use (10.6) to get

$$\dot{g}_K(t) = (1-\alpha)\big(g_A(t) + n - g_K(t)\big)g_K(t), \tag{10.10}$$

$$\dot{g}_A(t) = \big(\beta g_K(t) + \gamma n + (\theta - 1)g_A(t)\big)g_A(t), \tag{10.11}$$

which is a nonlinear dynamical system (a system of two first-order nonlinear differential equations) that describes the interaction between capital accumulation g_K and knowledge accumulation g_A. We shall proceed to analyse the dynamics of this system.

10.3 ANALYSIS, SIMULATION AND VISUALISATION

The steady-state or equilibrium points of the system (10.10)–(10.11) are found by setting both the equations to zero, and thus

$$\dot{g}_K(t) = 0, \quad \text{and} \quad \dot{g}_A(t) = 0. \tag{10.12}$$

First, $\dot{g}_K(t) = 0$ yields

$$(1-\alpha)\big(g_A^* + n - g_K^*\big)g_K^* = 0, \tag{10.13}$$

which in turn gives rise to two equilibrium values of g_K given by

$$g_K^* = 0 \quad \text{and} \quad g_K^* = g_A^* + n. \tag{10.14}$$

Similarly, setting $\dot{g}_A(t) = 0$ we get

$$\big(\beta g_K^* + \gamma n + (\theta - 1)g_A^*\big)g_A^* = 0, \tag{10.15}$$

which yields two equilibrium values of g_A given by

$$g_A^* = 0 \quad \text{and} \quad \beta g_K^* + \gamma n + (\theta - 1) g_A^* = 0. \tag{10.16}$$

Therefore, from (10.14) and (10.16), we find that the system (10.10)–(10.11) has four equilibrium points (g_K^*, g_A^*) that are given by

$$e_1^* = (0,0), \quad e_2^* = (n,0), \quad e_3^* = \left(0, \frac{\gamma n}{1-\theta}\right) \tag{10.17}$$

and

$$e_4^* = \left(\frac{n(\beta+\gamma)}{1-(\beta+\theta)} + n, \frac{n(\beta+\gamma)}{1-(\beta+\theta)}\right). \tag{10.18}$$

In order for the equilibrium points to make economic sense, we need to check the implied parametric conditions under which they are positive. We see that the equilibrium point $\left(g_K^*, g_A^*\right) = \left(0, \frac{\gamma n}{1-\theta}\right)$ is positive for $\theta < 1$ and negative for $\theta > 1$. In this case, from an economic point of view, we need $g_A > 0$, which implies the restriction that $\theta < 1$. Similarly, the equilibrium point $\left(g_K^*, g_A^*\right) = \left(\frac{n(1+\gamma-\theta)}{1-(\beta+\theta)}, \frac{n(\beta+\gamma)}{1-(\beta+\theta)}\right)$ only makes economic sense if $\beta + \theta < 1$. Note

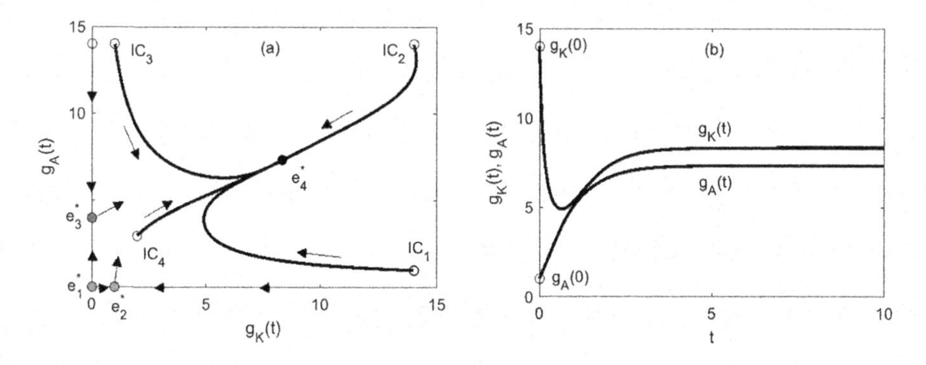

Figure 10.1: The endogenous growth model for the parameter values $\alpha = 0.5$, $\beta = 0.2$, $\gamma = 2$, $\theta = 0.5$ and $n = 1$. (a) A state-space plot showing four trajectories of $(g_K(t), g_A(t))$, for four different initial conditions, that all approach the stable equilibrium point e_4^*. The values of the equilibrium points are $e_1^* = (0,0)$, $e_2^* = (1,0)$, $e_3^* = (0,4)$ and $e_4^* = (8.33, 7.33)$. (b) Time-histories of $g_K(t)$ and $g_A(t)$) corresponding to the initial condition IC$_1$: (14,1) in (a). The three other initial conditions in (a) are at IC$_2$: (14,14), IC$_3$: (1,14) and IC$_4$: (2,3).

in particular that this point (e_4^*) is the only stable equilibrium point, of the four equilibrium points, where both g_K^* and g_A^* are non-zero and is thus of primary interest in this general model of interaction between capital accumulation and knowledge accumulation.

10.3.1 STABILITY ANALYSIS

To determine the local stability properties of the equilibrium points (g_K^*, g_A^*) (given in (10.17)-(10.18)) of the system (10.10)-(10.11) we can follow the method in Chapters 5 and 8 and calculate the eigenvalues of the corresponding Jacobian J, and evaluate it at each of the equilibrium points as given below.

$$J\left(g_K^*, g_A^*\right) = \begin{pmatrix} (1-\alpha)\left(g_A^* + n - 2g_K^*\right) & (1-\alpha)g_K^* \\ \beta g_A^* & \beta g_K^* + \gamma n + 2(\theta - 1)g_A^* \end{pmatrix}. \quad (10.19)$$

The Jacobian is far from trivial for analytical determination of the signs of the eigenvalues for all four equilibrium points. However for the three of them, (0,0), (n,0) and $\left(0, \frac{\gamma n}{1-\theta}\right)$, it is pretty straightforward. First, for $\left(g_K^*, g_A^*\right) = (0,0)$ we get

$$J(0,0) = \begin{pmatrix} n(1-\alpha) & 0 \\ 0 & \gamma_n \end{pmatrix}, \quad (10.20)$$

which has the eigenvalues $\mu_1 = n(1 - \alpha)$ and $\mu_2 = \gamma n$ that are both positive for $n > 0$, $0 < \alpha < 1$ and $\gamma > 0$ and thus the equilibrium point (0, 0) is an unstable or *repellor* node.

Second, evaluating the Jacobian at the second equilibrium $\left(g_K^*, g_A^*\right) = (n,0)$ yields

$$J(n,0) = \begin{pmatrix} -n(1-\alpha) & n(1-\alpha) \\ 0 & n(\beta+\gamma) \end{pmatrix}, \quad (10.21)$$

which has the eigenvalues $\mu_1 = -n(1 - \alpha)$ and $\mu_2 = (\beta + \gamma)n$. Since μ_1 is negative, given $n > 0$ and $0 < \alpha < 1$, and μ_2 is positive given $\beta > 0$ and $\gamma > 0$, the equilibrium point (n,0) is a saddle node, which is stable in one direction and unstable in the other direction (see Figure 10.1)

Third, evaluating the Jacobian at the third equilibrium point $\left(g_K^*, g_A^*\right) = \left(0, -\frac{\gamma n}{\theta - 1}\right)$ yields

$$J\left(0,-\frac{\gamma n}{\theta-1}\right)=\begin{pmatrix}(1-\alpha)n\left(\dfrac{\theta-1-\gamma}{\theta-1}\right) & 0\\ -\dfrac{\beta\gamma n}{\theta-1} & -\gamma_n\end{pmatrix}, \tag{10.22}$$

which has the eigenvalues $\mu_1=(1-\alpha)n\left(\frac{\theta-1-\gamma}{\theta-1}\right)$ and $\mu_2 = -\gamma n$ where μ_1 is negative if $0 < \theta - 1 < \gamma$ and positive if $\theta - 1 < 0 < \gamma$ or $0< \gamma < \theta - 1$ for $0 < \alpha < 1$, $n > 0$, $\gamma > 0$, and μ_2 is always negative. Thus, the equilibrium point $\left(0,\frac{-\gamma n}{\theta-1}\right)$ is stable (stable node) if $0 < \theta - 1 < \gamma$ and unstable (saddle node) if $\theta - 1 < 0 < \gamma$ or $0 < \gamma < \theta - 1$. This gives that the equilibrium point e_3^* is also a saddle point in the (g_K, g_A) plane.

Finally, a similar stability analysis can be done by evaluating the the Jacobian at the fourth equilibrium point $\left(g_K^*,g_A^*\right)=\left(g_A^*+n,g_A^*\right)=\left(\frac{n(1+\gamma-\theta)}{1-(\beta+\theta)},\frac{n(\beta+\gamma)}{1-(\beta+\theta)}\right)$, which is given by

$$J\left(\frac{n(\beta+\gamma)}{1-(\beta+\theta)}+n,\frac{n(\beta+\gamma)}{1-(\beta+\theta)}\right). \tag{10.23}$$

However, because of the complexity of parametric conditions involved in this case, instead of performing an analytical analysis we will numerically locate the eigenvalues at the same time as we simulate the system.

10.3.2 SIMULATION ANALYSIS AND VISUALISATION

To simulate and visualise the endogenous growth model we will follow the methodology that we used for the IS/LM model in Chapter 5 and the Dornbusch overshooting model in Chapter 8. The MATLAB® code to simulate the system of two first-order differential equations (10.10)–(10.11) is given below in five separate parts, namely, (i) the initiation of the simulation and parameter values, (ii) the initial conditions, the call of the differential equation solver ode45 and the output, (iii) the endogenous growth model system, (iv) calculation of the eigenvalues and eigenvectors, and (v) plotting of the trajectories $(g_K(t), g_A(t))$ and time histories of $g_K(t)$ and $g_A(t)$, as seen in Figure 10.1. The first set of code clears the memory, closes all the figures, sets the parameters and calculates the equilibrium points and builds the Jacobian.

Initiation & parameter values & Jacobian & eigenvalues

```
%Endogenous growth model
clear all
close all
format compact

%Parameters
alpha  = 0.5;
beta   = 0.2;
gamma  = 2;
theta  = 0.5;
n      = 1;

params = [alpha, beta, gamma, theta, n];

% Jacobian
J4 = [(1-alpha)*(eq4(2)+n-2*eq4(1)),(1-alpha)*eq4(1);...
      beta*eq4(2),beta*eq4(1)+gamma*n+2*(theta-1)*eq4(2)];

% Equilibrium points
eq1 = [0,0];
eq2 = [n,0];
eq3 = [0,gamma*n/(1-theta)];
eq4 = [n*(beta+gamma)/(1-(beta+theta))+n,...
       n*(beta+gamma)/(1-(beta+theta))];
```

Next we set up the simulation and decide simulation time, initial conditions, solver properties (as in Section 5.3). The output from the ODE solver is a time vector t and 2-column state matrix Y, where the first column is $g_K(t)$ and the second column is $g_K(t)$ and thus we separate the output in a g_K and g_A vector that will be used for plotting the figures.

Initial conditions and the ODE solver

```
%Initial conditions
g_K0 = 14;
g_A0 = 1;
y0 = [g_K0;g_A0];
```

```
%Simulation time
t_start = 0;
t_end   = 10;

% Simulation setup
options = odeset('RelTol',1e-6,'AbsTol',1e-6);

%Simulation
[t,Y] = ode45(@(t,y) endogenous_system(t,y,params),...
        [t_start t_end],y0,options);

%Output
g_K = Y(:,1);
g_A = Y(:,2);
```

The ODE solver calls the function `endogenous_system(t,y,params)`, which evaluates the right-hand side of (10.10)–(10.11), with time (`t`), state variables (`y`), and parameter vector (`params`) as input.

The Endogenous growth model system

```
function dy = endogenous_system(t,y,params)

%Parameters
alpha = params(1);
beta  = params(2);
gamma = params(3);
theta = params(4);
n     = params(5);

%Variables
g_K = y(1);
g_A = y(2);

%The ODE model
dg_K = (1-alpha)*(g_A+n-g_K)*g_K;
dg_A = (beta*g_K+gamma*n + (theta-1)*g_A)*g_A;

dy = [dg_K;dg_A];

end
```

In order to calculate the eigenvalues of the Jacobian J, which was discussed above, we use the MATLAB® code

Eigenvalues

```
%Eigenvalues and eigenvectors
[eigenvectors4,eigenvalues4] = eig(J4);
mu4_1 = eigenvalues4(1,1)
mu4_2 = eigenvalues4(2,2)
```

that gives the output

```
mu4_1 =
   -6.4013
mu4_2 =
   -1.4320
```

Since we are focused on the fourth equilibrium point of (10.18) here, we see that both eigenvalues are negative, which indicates that the equilibrium point is a stable node.

To visualise the output of the simulation, we follow the same methodology as in the previous chapters.

Visualisation

```
%Plotting
figure(1), hold on, box on
plot(t,g_K,'k','LineWidth',2)
plot(t,g_A,'k','LineWidth',2)
xlabel('t'), ylabel('g_K(t), g_A(t)')

figure(2), hold on, box on
plot(g_K,g_A,'k','LineWidth',2)
plot(eq1(1),eq1(2),'o','MarkerFaceColor','k',...
    'MarkerEdgeColor','k')
plot(eq2(1),eq2(2),'o','MarkerFaceColor','k',...
    'MarkerEdgeColor','k')
plot(eq3(1),eq3(2),'o','MarkerFaceColor','k',...
    'MarkerEdgeColor','k')
plot(eq4(1),eq4(2),'o','MarkerFaceColor','k',...
    'MarkerEdgeColor','k')
xlabel('g_K(t)'), ylabel('g_A(t)')
```

10.3.3 DISCUSSION

The analysis of the dynamic interaction between capital accumulation g_K and knowledge accumulation g_A as described by equations (10.10)–(10.11) yields some interesting results. The dynamics of the system is visualised in Figure 10.1 and a few observations are in order. First, the four equilibrium points of the system (10.18) are denoted by e_1^*, e_2^*, e_3^* and e_4^* in the $(g_K(t),g_A(t))$ plane. As shown in the stability analysis section, the trivial equilibrium point e_1^* at the origin is an unstable, or a *repellor* node. This is visualised by the arrows indicating the direction of the vector field, thus pointing outwards from the equilibrium point. Second, the equilibrium point e_2^* is a saddle node – as can be seen from the arrows it is stable in one direction (arrows going towards the node), i.e. along the $g_K(t)$ axis and unstable in the vertical direction (arrows going away from the node), i.e. along the $g_A(t)$ axis. In fact, the equilibrium point e_2^*, where $g_K^* = n$ and $g_A^* = 0$, yields the result of the original Solow model, where growth in capital accumulation converges to the stead state rate n, here, and permanent growth can only be achieved by increasing the technology parameter A. However, when we introduce the dynamics of knowledge accumulation and study its interaction with the capital accumulation as in (10.10)–(10.11), the generalised Solow model in this 2-dimensional set-up yields a saddle node equilibrium, which is stable along the capital accumulation axis $g_K(t)$ and unstable along the knowledge accumulation axis $g_A(t)$.

Similarly, the equilibrium point e_3^* yields the result of another special case, where only the dynamics of knowledge accumulation $\dot{g}_A(t)$ is considered without the capital accumulation $\dot{g}_K(t)$. In this special case, the equilibrium point e_3^* is a stable node, with arrows pointing inwards from both directions along the $g_A(t)$ axis. However, when we introduce the dynamics of capital accumulation and study its interaction with the knowledge accumulation, the general model (10.10)–(10.11) yields a saddle node dynamics at e_3^*. It is stable along the $g_A(t)$ axis, as shown by arrows pointing towards the e_3^*, and unstable along the $g_k(t)$ axis shown by arrows pointing outwards the equilibrium point e_3^*.

The only stable equilibrium point in this system (10.10)–(10.11) is e_4^*, as shown in Figure 10.1, i.e. it is the only *attractor* in the $(g_K(t), g_A(t))$ plane. As visualised in Figure 10.1, every point in the plane is attracted towards the equilibrium point as shown by the arrows. Here we have shown dynamic behaviour for four initial conditions IC_1, IC_2, IC_3 and IC_4 – four different points of $(g_K(0), g_A(0))$ – and we can see they all converge to the equilibrium point e_4^*. Think of the initial conditions as countries that are similar with the same parameter values (α, β, γ and θ) in their production functions, but start at different initial conditions. For instance, the country starting from the initial condition IC_1 has a very high growth in capital accumulation but a low growth in knowledge accumulation. The trajectory of convergence reveals that as the interaction between knowledge accumulation and capital accumulation ensues, the rate of growth in capital accumulation slows down initially as the growth in knowledge accumulation increases and then both increase in tandem to reach the equilibrium growth rate e_4^*. This could be thought of as a case where the growth in knowledge accumulation begins to rationalise and reshape the process of capital accumulation where there is an initial reduction in the rate of growth in the latter and then an increase to reach the steady state growth. In other words, the out of equilibrium dynamics could be thought of as a Schumpeterian process of *creative destruction,* where the tangible capital stock is being creatively destroyed to produce new efficient, or smarter, machines by the process of knowledge accumulation, and together they generate growth and converge the economy to the steady state growth rate e_4^*. The time histories in Figure 10.1(b) show the convergence of the $g_K(t)$ and $g_A(t)$ to the steady growth e_4^* for the initial condition IC_1.

Similarly, an economy starting with a high knowledge accumulation and low capital accumulation, such as at IC_3, also shows a non-linear path of reaching the steady state growth. Further, when an economy starts with a high (or low) capital and knowledge accumulation, for the initial condition at IC_2, (or IC_4), it takes a more direct path to reach the steady state growth. While these out of equilibrium dynamics are subject to the particular parameter values of the respective production functions, it would be interesting to calibrate the model for different economies and study whether the model could explain the historical experience of particular economies.

10.4 CONCLUDING REMARKS, EXTENSIONS AND CHALLENGES

In this chapter, we explored the dynamics of interaction between capital accumulation and knowledge accumulation in a general model. The analysis revealed that the Solow model and the pure knowledge accumulation based growth models emerge as special cases in the general set up. Moreover, we also highlighted that in the model the stability properties of the special cases change from stable nodes, or yielding steady growth, to unstable (saddle) nodes along one direction either in $g_A(t)$ (knowledge accumulation) or $g_K(t)$

(capital accumulation) axes when the interaction between them is explicitly considered. The out of equilibrium dynamics showed interesting paths of convergence, albeit for the particular numerical parameter values considered here. These paths of convergence could be of interest for policy analysis. For instance, the convergence path of the initial condition IC_1 in Figure 10.1(a) may prompt questions about investment in the type of knowledge that can generate growth in the economy, which starts with a historically high level of tangible capital.

In this model, and in the family of such models, knowledge accumulation is treated as a homogeneous entity like the tangible capital in the production function – a fraction of labour a_L and capital a_K in the R&D sector produce the output called 'knowledge'. But knowledge is not homogeneous as it comes in different forms such as the basic theoretical to the highly applied knowledge. There is also knowledge about specific goods, that may not apply outside that particular species of goods. Furthermore, although all knowledge is in principle *nonrival,* they can be excluded from general use depending on the type of property rights regime. The degree of property rights on knowledge can have an impact on the knowledge production and consequently on economic growth. In the Covid-19 context, the vaccine recipes are protected by the intellectual property rights regime, which had its impact both on the vaccination programme in different countries and the consequent differential impact on economic recovery around the world. Within this broad institutional context, the endogenous growth literature concentrates on the determinants of knowledge accumulation. Some of these include creating private incentives for R&D, the development of entrepreneurship and enabling environment for learning by doing, among others.

The central motivation of endogenous growth models, as in the case of Solow's growth model, is to explain variations in long-run growth and also cross-country growth differentials. Arising from the theoretical literature, there is a huge empirical literature exploring whether economic growth is driven by the factors proposed by these models.

Allied to these explorations there is also the question of what determines technological progress. The endogenous growth models suggest that technological progress is an increasing function of population size.

The intuition behind the argument is simple: the larger the population, the more people available to make discoveries resulting in rapid knowledge accumulation leading to growth. The impact of endogenous growth models can be seen in recent policy focus, both in developing and developed countries, in terms of creating knowledge-based economies and underscoring the importance of skill development. However, while these proximate determinants of knowledge accumulation and knowledge-based economies are well articulated in the endogenous growth literature, the social structure underpinning these determinants seems to be assumed rather than explained or included in these models. Further extensions to include these fundamental determinants in the knowledge accumulation or human capital models would not only enrich the literature but also be insightful and impactful from a policy perspective.

NOTES

1 Romer's introduction of nonrival ideas provides a solution to the *Maluthusian problem* in Solow's model where an increase in population size adversely impacts the steady per capita growth. In Romer's conception, the increase in population size need not put brakes on economic growth as long as it is offset by the growth in the stock knowledge (A).

2 For pedagogic purposes, we consider the standard representation as in David Romer [6].

BIBLIOGRAPHY

[1] Abramovitz, M. (1956), Resource and Output Trends in the United States since 1870, *American Economic Review,* Vol. 46, pp. 5–23.

[2] Lucas jr., R.E. (1988), On the Mechanics of Economic Development, *Journal of Monetary Economics,* Vol. 22, pp. 3–42.

[3] Lucas jr., R.E. (1990), Why Doesn't Capital Flow from Rich to Poor Countries?, *American Economic Review,* Vol. 80, pp. 92–96.

[4] Mankiw, N.G., Romer, D. and Weil, D.N. (1992), A Contribution to the Empirics of Economic Growth, *The Quarterly Journal of Economics,* Vol. 107, No. 2, pp. 407–437.

[5] Jones, C. (2002), Sources of U.S. Economic Growth in a World of Ideas, *American Economic Review*, Vol. 92, pp. 220–239.

[6] Romer, D. (2004), *Advanced Macroeconomics 4th edition,* Douglas Reiner.

[7] Romer, P.M. (1986), Increasing Returns and Long Run Growth, *Journal of Political Economy,* Vol. 94, pp. 1002–1037.
[8] Romer, P.M. (1990), Endogenous Technological Change, *Journal of Political Economy,* Vol. 98, pp. 71–102.
[9] Solow, R.M. (1957), Technical Change and the Aggregate Production Function, *Review of Economics and Statistics,* Vol. 39, pp. 312–320.

11

BUSINESS CYCLES I: SAMUELSON'S MULTIPLIER-ACCELERATOR MODEL

One of the main objectives of the study of macroeconomics is to understand the business cycle fluctuations. While the classical economists were interested in long cycles in the capitalist economies, the early contributors in the Keynesian tradition such as Paul Samuelson, John Hicks and others like Michal Kalecki, who straddled between the two traditions, were interested in understanding the cyclical fluctuations in the short run. The business cycle studies were at the heart of macroeconomics research in the 1920s and the 1930s. Paul Samuelson's paper published in 1939 is one of the seminal contributions to the business cycle literature. The publication of the *multiplier-accelerator model* [11] in many ways laid out a rigorous dynamical systems approach formalism to the *theory of business cycles*. Note that the analysis of the business cycle in the Keynesian tradition considers the aggregate structure of the economy and investigates the fluctuations in output in the aggregate variables such as investment, consumption, or government expenditure etc. The modern business cycle studies, which are interestingly labelled as the *new Keynesian models*, are articulated using the micro-foundations approach that locates the origin of fluctuations in the economic decisions of rational agents in the economy. In Chapter 12 we will study one such model called the *real business cycle model*.

In terms of the Keynesian tradition in the business cycle studies, the formalism adopted in Samuelson's article [11], though limited in terms of the dynamics generated by the model, led to the subsequent theoretical developments in this area. Samuelson himself was aware of the limitation of his model, which he explicitly stated

> . . .this representation is strictly a *marginal* analysis to be applied to the study of small oscillations. . .
>
> (Samuelson [11], p. 78.)

The implication of such a marginal analysis is that the time series generated by any particular solution of the model will determine actual income only

DOI: 10.4324/9780429324406-11

for a short time about a market equilibrium. The limitation in the possible dynamics produced by the model, in particular the case of generating sustained oscillations, led to a great body of work by way of extensions and modifications of the original Samuelson's model that continues to evolve to the present day.

The initial developments came from Hicks [6], Goodwin [5] and Kaldor [7]. Both Hicks and Goodwin articulated models that generated non-damped cycles. Hicks's trade-cycle model is a non-linear modification of Samuelson's multiplier-accelerator model with an upper bound on the income as a 'ceiling' and a 'floor' set by the depreciation rate of investment. More precisely, Hicks rewrites the induced part of the investment equation in Samuelson's model as piecewise-linear equations. The economic intuition behind the piecewise-linear transformation is that as the income increases, the induced part of the investment becomes positive and reinforces the rise in income and the upswing of the economy may continue until the income reaches its ceiling level. When income remains at the ceiling level, the induced investment becomes zero causing a down swing of the economy. Moreover, with the assumption of the autonomous investment growing exogenously at a constant rate, Hicks was able to show the possibility of how a growth process can be coupled with the business cycles.[1] Goodwin, in his pioneering work [5], endogenised the Hicksian ceiling and floor as being determined by the desired capital stock and consequently the discontinuities in net investment generates sustained oscillations in the rate of change in both the level of output and the actual capital stock. However, later in the paper, by introducing lags in the accelerator and multiplier coefficients Goodwin shows that his non-linear flexible accelerator model is capable of generating self-sustained fluctuations [9].

In more recent times, with the hindsight of developments in dynamical systems theory, the issue of generating sustained oscillations in the class of multiplier-accelerator models was reconsidered by Hommes [4], Gallegati *et al.* [2] and Puu *et al.* [10]. Hommes revisits this issue from the point of view of periodicity in the dynamic behavior generated by Hicks' model. In particular, Hommes poses the question whether every time path in Hicks' trade cycle model converges to a periodic time path or not. He extends Hick's model by considering lags in consumption and/or investment, i.e. being distributed over several time periods, and demonstrates the existence of quasi-periodic orbits and strange attractors. From the perspective of the dynamics, Hommes' extension of Hick's trade-cycle model is interesting in that the attractors in the model exhibit periodic behaviour interspersed by sudden bursts of erratic behaviour, which is pertinent for understanding regime shifts that we encounter in real economies. In a more detailed investigation of the dynamics of Hicks' model, Gallegati *et al.* [2] use bifurcation analysis to study the conditions under which the model produces periodic and quasi-periodic dynamics. In particular, their study reveals the nature of the attracting set, for the values of the accelerator coefficient where Hicks' original model turns unstable. In another interesting paper on the reformulation of Hicks' trade cycle model, Puu *et al.* [10] revisit the original issue of growth oscillations using the relative deviations approach. In order to generate growth and cycles in the original Hicks model, Puu *et al.* [10] relate the 'floor' directly to the growth in capital stock through a constant depreciation rate with the rationale that as capital stock grows it increases the absolute value of maximum disinvestment and hence the 'floor' should be falling or decreasing with capital accumulation. With this set-up, the authors reformulate Hicks' model to generate endogenous growth trend and show that the model can produce cycles of various periodicities.

From this brief review of the literature, it can be seen that most of these extensions in the class of multiplier-accelerator models followed Samuelson-Hicks or Goodwin frameworks to generate sustained fluctuations. Since our aim in this chapter is to introduce you to this approach to business cycle modelling, we use Samuelson's model, which is in a way relatively straightforward, in its original form, and study the range of dynamics generated by the model.

11.1 MODELLING

The Samuelson multiplier-accelerator model is a discrete model of a closed economy.

The main aim of the model is to understand the fluctuations in the level of output through the dynamic interaction between the *multiplier* and the *accelerator*. The set up of the model is as follows:

$$C_t = bY_{t-1}, \quad 0 < b < 1, \tag{11.1}$$

$$I_t = I_t^g + I_t^p, \tag{11.2}$$

$$I_t^g = G, \quad G > 0, \tag{11.3}$$

$$I_t^p = k\left(C_t - C_{t-1}\right), \quad k > 0, \tag{11.4}$$

$$Y_t = C_t + I_t. \tag{11.5}$$

where Y_t denotes income (national income), C_t denotes consumption, and I_t stands for total investment. Note that the model is a discrete model with time periods denoted by $t = 0,1,2,\ldots$

Let us now understand the equations of the model. First, the aggregate consumption expenditure (11.1) is a function of the previous period's income Y_{t-1}, where the coefficient b is the *marginal propensity to consume* (MPC). Second, (11.2) is the *investment function* that has two terms, I^g and I^p, where the former stands for autonomous investment, which is assumed to be a constant (G) as given by equation (11.3), and the latter is the induced investment as in (11.4), where k is the *accelerator coefficient*. It is assumed that induced investment is driven by consumer demand and increases as the consumer spending C increases between t and t-1. Finally, (11.5) represents the macroeconomic income-expenditure equilibrium condition, or the savings-investment equilibrium condition.

Substituting (11.1), (11.3), (11.4), in (11.2) yields,

$$I_t = G + k\left(C_t - C_{t-1}\right) = kb\left(Y_{t-1} - Y_{t-2}\right) + G, \tag{11.6}$$

which together with (11.5) implies

$$Y_t = C_t + I_t = bY_{t-1} + bkY_{t-1} - bkY_{t-2} + G. \tag{11.7}$$

and rewriting (11.7) gives

$$Y_t - b(1+k)Y_{t-1} + bkY_{t-2} = G. \tag{11.8}$$

Equation (11.8), a second-order non-homogeneous difference equation, is the fundamental equation of Samuelson's multiplier-accelerator model. We note that are only two parameters in the model, namely, the marginal propensity to consume (MPC) parameter b and the accelerator coefficient k, determine the dynamics of the model. The analysis of the model will reveal how the interaction between the multiplier b and the accelerator k generates various dynamics in output (Y_t).

11.2 ANALYSIS, SIMULATION AND VISUALISATION

The Samuelson model is relatively easy to solve, but it leads to non-trivial solutions that are dependent on the relationship between b and k. We will go through the solution to explain this point. First, we know from theory of difference equations (see Appendix A.2) that the solution is given by the sum of the *homogeneous* $\left(Y_t^h\right)$ and *particular* $\left(Y_t^p\right)$ *solutions* so that

$$Y_t = Y_t^h + Y_t^p, \quad t = 0, 1, \ldots. \tag{11.9}$$

Since the right-hand side of (11.8) is assumed to be constant, we can easily determine that the particular solution Y_t^p is the same as the fixed value Y^* of the output Y_t, which is given by

$$Y_t^p = Y^* = \frac{G}{1-b}. \tag{11.10}$$

To determine the homogeneous solution we need to solve the *homogeneous equation* given by

$$Y_t^h - b(1+k)Y_{t-1}^h + bkY_{t-2}^h = 0, \tag{11.11}$$

whose characteristic equation

$$\mu^2 - b(1+k)\mu + bk = 0 \tag{11.12}$$

gives rise to the roots

$$\mu_1 = \frac{b(1+k) + \sqrt{b^2(1+k)^2 - 4bk}}{2}, \quad \mu_2 = \frac{b(1+k) - \sqrt{b^2(1+k)^2 - 4bk}}{2}.$$

Now, depending on the sign of the discriminant

$$\Delta = b^2(1+k)^2 - 4bk,$$

μ_1 and μ_2 can either be real or imaginary, and thus leading to different characteristics to the solution of the homogeneous equation, such that

$$\Delta > 0: \quad Y_t^h = c_1\mu_1^t + c_2\mu_2^t, \tag{11.13}$$

$$\Delta < 0: \quad Y_t^h = r^t\left(c_1\sin(\theta t) + c_2\cos(\theta t)\right) \tag{11.14}$$

for $t = 0, 1, \ldots$, where

$$r = |\mu_i| \quad \text{and} \quad \theta = \arg(\mu_i) = \tan^{-1}\left(\frac{\mathrm{Im}(\mu_i)}{\mathrm{Re}(\mu_i)}\right), \quad i = 1, 2,$$

and the parameters c_1 and c_2 are given by the initial conditions. Therefore, the solution to (11.8) can be summarised as

$$Y_t = \frac{G}{1-b} + \begin{cases} c_1\mu_1^t + c_2\mu_2^t, & \Delta > 0, \\ r^t\left(c_1 \sin(\theta t) + c_2 \cos(\theta t)\right), & \Delta > 0. \end{cases} \tag{11.15}$$

The necessary condition for the convergence to the fixed point or the equilibrium value of output (Y^*), such that the homogeneous solution $Y_t^H \to 0$, is $bk < 1$ (see Appendix A.2 for the analytical proof). Further analysis shows that this condition is in fact is sufficient to guarantee convergence to the equilibrium, i.e.

$$bk < 1 \Rightarrow Y_t \to Y^* = \frac{G}{1-b} (\text{convergence}) \tag{11.16}$$

$$bk > 1 \Rightarrow |Y_t| \to \infty (\text{divergence}) \tag{11.17}$$

However, the sign of the discriminant Δ determines whether the solution is oscillatory or monotonic. In particular, we note that

$$\Delta = b^2(1+k)^2 - (4bk) < 0 \quad \Leftrightarrow \quad b < \frac{4k}{(1+k)^2} \quad \text{(oscillatory)}, \tag{11.18}$$

$$\Delta = b^2(1+k)^2 - (4bk) > 0 \quad \Leftrightarrow \quad b > \frac{4k}{(1+k)^2} \quad \text{(monotonic)}. \tag{11.19}$$

We can graphically describe the stability properties of Y^* in terms of four regions in the parameter space (k, b) for $0 < b < 1$ and $k > 0$, determined by the inequalities (11.18) and (11.19). Furthermore, the boundaries of these regions are given by the identities

$$b = \frac{4k}{(1+k)^2}, \quad b = \frac{1}{k}, \quad b = 0 \quad \text{and} \quad b = 1.$$

In Figure 11.1 we see four regions A–D defined by the four identities (two lines and two curves) that also define the dynamics about Y^*. In region A, the roots μ_1 and μ_2 of the characteristic equation (11.12) are real, positive and less than 1, and the dynamics monotonically converges to the equilibrium value Y^*. In region B the roots μ_1 and μ_2 are complex and less than 1 in magnitude and thus generate oscillatory convergence to Y^*. In region C, the roots μ_1 and μ_2 are complex and greater than 1 in magnitude and thus there is a oscillatory divergence from Y^*. In region D, the roots $\mu_1 > \mu_2$ are real, positive and at least μ_1 is greater than 1 and thus the system exhibits monotonic divergence from Y^*.

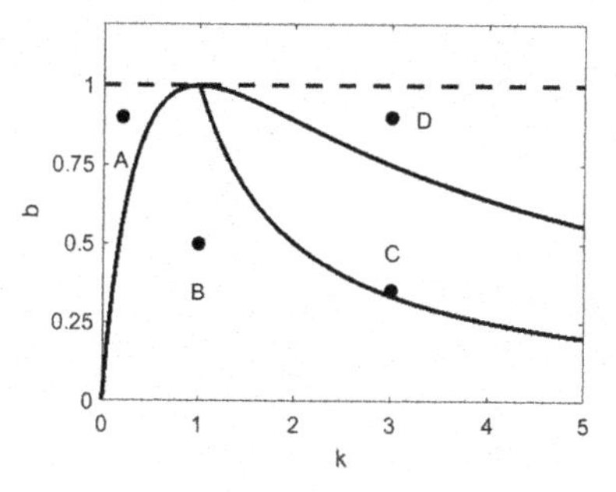

Figure 11.1: Samuelson's original qualitative stability diagram

We can further analyse the dynamics in the different regions from the time histories of the output Y_t in Figure 11.2. In region A the system

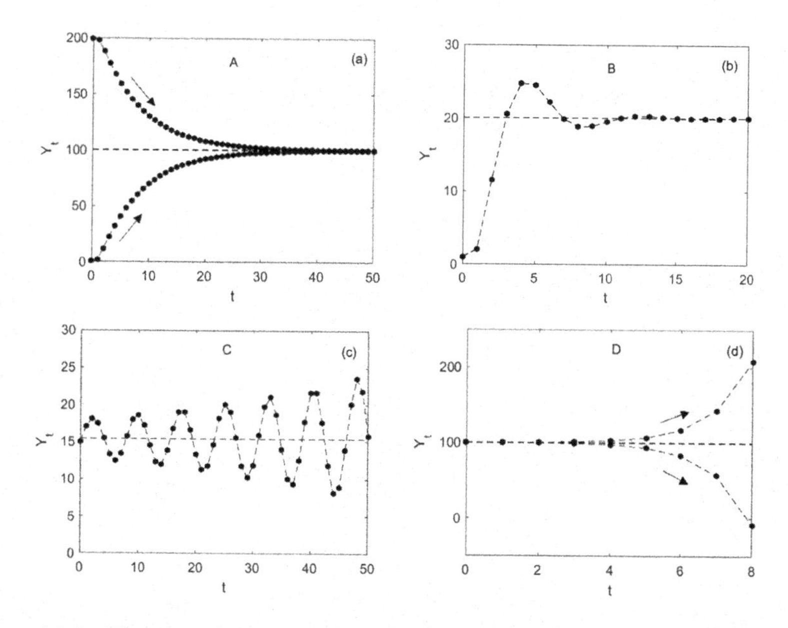

Figure 11.2: Times-series corresponding to regions A–D in the (k,b)-parameter space in Figure 11.1. The parameter values are G and in (a) $(k,b) = (0.2,0.9)$, (b) $(k,b) = (1,0.5)$, (c) $(k,b) = (3,0.35)$ and (d) $(k,b) = (3,0.9)$. In (a) two different initial conditions converge monotonically to the equilibrium value, in (b) the dynamics of the trajectory shows oscillatory convergence to the equilibrium, in (c) the dynamics generates oscillatory divergence away from the equilibrium, and in (d) the dynamics yields monotonic divergence away from the equilibrium as shown by the time-series of two different initial conditions diverging monotonically.

exhibits monotonic convergence and thus starting from any initial condition output converges monotonically to its equilibrium value $\left(y_t^*\right)$. In region B, the dynamics of output shows oscillatory convergence, in region C oscillatory divergence and in region D monotonic divergence. The dynamics in the respective regions of Figure 11.1 correspond to the time-series in Figures 11.2 (a)–(d), respectively.

We can now understand the economic intuition behind the dynamics of the model. For any given value of the MPC parameter b, as the acceleration coefficient k is increased the model turns from a stable system, showing monotonic and oscillatory convergence to the equilibrium, to an unstable system exhibiting oscillatory and monotonic divergence away from the equilibrium. In particular, when the acceleration coefficient remains below 1 the system always exhibits stability as given by stability condition ($bk < 1$). However, when the acceleration coefficient exceeds 1, the system loses stability ($bk > 1$) as the dynamics of the system is taken over by the stronger acceleration effect, where firms will respond to higher demand and increase their investment, which in turn leads to higher output (demand) via the multiplier, and the positive feedback between the multiplier and accelerator takes the system away from the equilibrium.

The MATLAB® code to generate the stability diagram in Figure 11.1 is given in the following.

Samuelson's stability diagram

```
%Preamble
clear all
close all

%Parameters
k_end = 5;

%Intial Conditions
k1 = linspace(0,k_end,100);
k2 = linspaced(1,k_end,100);
k3 = [0,k_end];
b1 = 4*k1./(1+kl).^2;
b2 = 1./k2;
b3 = [1,1]
```

```
%Plotting
figure(9),hold on
plot(k1,b1,'k','LineWidth',2')
plot(k2,b2,'k','LineWidth',2')
plot(k3,b3,'k--','LineWidth',2')
xlabel('k'),ylabel('b')
box on, hold off
axis([0 5 0 1.2])
```

Simulating the Samuelson model (11.8) follows the same method that we used for other difference equations in Chapters 2 and 3.

Simulation of Samuelson's model

```
%Preamble
clear all
close all

%Parameters
b = 0.9; %0 < b < 1
k = 0.2; %k > 0
G = 10;

%Intial Conditions
Y0 = G/(1-b)-0.1;
Y1 = G/(1-b)-0.2;

%Simulation parameters
n = 8; % Number of iterations

%Simulation
Y(1) = Y0;
Y(2) = Y1;
for t = 3:n+1
  Y(t) = b*(1+k)*Y(t-1) - b*k*Y(t-2) + G;
end

%Fixed point
Yfp = G/(1-b);
```

```
%Plotting
figure(49),hold on
plot(0:n,Y,'ko','MarkerFaceColor','k','Markersize',4)
plot([0,n],[Yfp,Yfp],'k--')
xlabel('t'),ylabel('Y_t')
box on, hold off
```

Numerical explorations

Readers are encouraged to experiment with the parameter values and the initial conditions to try and visualise the dynamics of the model. For instance, using the MATLAB® codes above, you are encouraged to explore the parametric values for which the model generates period orbits. Hint: Explore the boundaries of the regions in Figure 11.1.

11.3 CONCLUDING REMARKS, EXTENSIONS AND CHALLENGES

Samuelson's multiplier-accelerator model laid a rigorous foundation for the study of business cycles. The model is set in the framework of Keynesian economics, which began flourishing in the 1930s, and articulates a dynamical-systems approach to the study of business cycles. As we noted earlier, initial extensions of the model came from Hicks [6] where he was able to show how a growth process can be coupled with the business cycles. The growth part in the Hicksian model was introduced via the assumption of exogenously growing autonomous investment in Samuelson's model. However, in order to generate non-damped cycles, Hicks introduced boundaries such as 'ceiling' and 'floor' within which the dynamics exhibit business cycles around the trend rate of growth. Goodwin [5] further extended Hick's concept of ceiling with the desired level of capital stock and was able to show self-sustained oscillations within the general framework of the multiplier-accelerator interaction.

The intellectual context of the initial wave of business cycle models was surely Keynesian in the sense that they were focused on variables such as investment, capital stock etc., in the aggregate. Looking at these models in the current intellectual context of the *new Keynesian economics*, the first generation business cycle models proposed by Samuelson, Hicks and Goodwin arguably lack the necessary micro-foundations. One of the implications of the *Lucas critique*[8] is that of articulating the necessary microeconomic behavioural foundations for the study of business cycles, and macroeconomics in general. The modern business cycle studies articulate such micro-founded business cycle models and in the next chapter, we will study the the foundational model in the genre called the *Real Business Cycle* (RBC) model.

NOTE

1 Here we provide a schematic overview pertaining to the dynamics of various models in this area. We would refer the interested reader to other excellent sources for a detailed discussion on the models. For an elaborate discussion on Hicks' model, see Goodwin [5], Dussenberry [1], Gandolfo [3] and Tu [12].

BIBLIOGRAPHY

[1] Duessenberry, J. (1950), Hicks on the Trade Cycle, *The Quarterly Journal of Economics*, Vol. 64, pp. 464–476.

[2] Gallegati, C., Gardini, L., Puu, T. and Sushko, I. (2003), Hicks' Trade Cycle Revisited: Cycles and Bifurcations, *Mathematics and Computers in Simulation*, Vol. 63, pp. 505–527.

[3] Gandolfo, G. (1997), *Economic Dynamics*, Springer-verlag, Berlin.

[4] Hommes, C. (1995), A Reconsideration of Hicks' Non-Linear Trade Cycle Model, *Structural Change and Economic Dynamics*, Vol. 6, pp. 435–459.

[5] Goodwin, R. (1951), The Nonlinear Accelerator and the Persistence of Business Cycles, *Econometrica*, Vol. 19, pp. 1–17.

[6] Hicks, J. (1950), *A Contribution to the Theory of Trade Cycle*, Clarendon Press: Oxford.

[7] Kaldor, N. (1940), A Model of the Trade Cycle, *Economic Journal*, Vol. 50, pp. 78–92.

[8] Lucas, R.E. (1976), Econometric Policy Evaluation: A Critique, In Brunner, K. and Metzler, A (eds.) *The Phillips Curve and Labour Markets Carnegie-Rochester Conference Series on Public Policy*, Vol. 1, New York: American Elsevier, pp. 19–46.

[9] Matsumoto, A. (2009), Note on Goodwin's 1951 Nonlinear Accelerator Model with an Investment Delay, *Journal of Economic Dynamics and Control*, Vol. 33, No. 4, pp. 832–842.
[10] Puu, T., Gardini, L. and Sushko, I. (2005), Hicksian Multiplier-accelerator Model with Floor determined by Capital Stock, *Journal of Economic Behaviour and Organization*, Vol. 56, pp. 331–348.
[11] Samuelson, P. (1939), Interaction between the Multiplier Analysis and the Principle of Acceleration, *The Review of Economic Statistics*, Vol. 20, pp. 75–78.
[12] Tu, P. (1992), *Dynamical Systems: An Introduction with Applications in Economics and Biology*, Springer-Verlag: Berlin.

12 BUSINESS CYCLES II: THE REAL BUSINESS CYCLE MODEL

12.1 ECONOMIC PROBLEM

The family of business cycle models can be divided into two broad groups. In one group we can assimilate all the models that consider the economy from an aggregate point of view and argue that the business cycle fluctuations are driven mainly by the volatility in the aggregate variables, such as the investment expenditure, due to a wide range of reasons from non-realisation of expectations to animal spirits. Broadly speaking, this group of aggregate economy models can be labelled as the Keynesian business cycle models with notable examples including Samuelson and Hick's Multiplier-Accelerator models seen in Chapter 11. The second group comprises those models that articulate micro-foundations to the aggregate economy and argue that the business cycle fluctuations arise as a result of the rational response of agents to unanticipated shocks to the economy. The most prominent and widely used model in this group is the *real business cycle model* (RBC), which is the subject of study in this chapter.

The real business cycle model form the basis for the more advanced *dynamic stochastic general equilibrium* (DSGE) model, which is the benchmark model for macroeconomic policy analysis. The real business cycle model is an extension of the conventional general equilibrium models and where the representative agent maximises life-time utility rather than utility at a point in time. These models are derived from the class of Ramsey-Cass-Koopmans inter-temporal optimal growth models (see the box below).

In the following, we shall explore the general analytical architecture that underpins real business cycle models and in turn discuss and analyse those building blocks in detail.

DOI: 10.4324/9780429324406-12

The Ramsey-Cass-Koopman growth models

The Ramsey-Cass-Koopman class of growth models generalises Solow's growth model by endogenising the savings rate and proposes the determination of the optimal growth in a decentralised market economy. These models followed Frank Ramsey's [6] pioneering articulation of the optimal saving model published in 1928 as a central planner's problem of maximizing social welfare over successive generations. David Cass [1] and Tjalling Koopmans [5] simultaneously and independently adopted and extended Ramsey's model to a decentralised economy setting.

12.2 MODELLING

In terms of the basic RBC model, we assume a private economy, with households and firms, that is closed for foreign trade and which lasts for T discrete time periods.[1] Let us further discuss various assumptions and set up the economic problem of the households and firms in this economy.

HOUSEHOLDS

There exists a large number of identical households and for simplicity the total number of households is normalised to 1. The fact that we assume a large number of identical households intuitively means that households are assumed to behave competitively, i.e. no individual household can affect the behaviour of prices in the economy. The households do not take leisure and they work 1 unit of time per period. In other words, households work all the time and leisure does not give them any utility. The economic choice that the households have to make in each period is how much to consume in the current period and how much to save for the future.

Let $c_t \geq 0$ denote the household's consumption and $u(c_t)$ its utility at time t, and assume that each household's *life-time utility function* U is of the form[2]

$$U(c_0, c_1, \ldots, c_T) = u(c_0) + \beta u(c_1) + \beta^2 u(c_2) + \cdots + \beta^T u(c_T)$$

$$= \sum_{t=0}^{T} \beta^t u(c_t), \tag{12.1}$$

where $\beta \in (0,1)$ is the *discount factor* that reflects the degree of impatience the household has for future consumption, i.e. it derives less utility from the same consumption if that consumption occurs at a future date.

Household impatience

In practice, the degree of household's impatience is measured by the *time discount rate* ρ that is related to the discount factor β through

$$\beta = \frac{1}{1+\rho}.$$

A high discount rate ρ implies a low discount factor β, which means that the households are more impatient towards future consumption.

Now, let us write down the inter-temporal budget constraint of the households. The households work one unit of time each period, for which they earn a wage income of w_t. They also earn interest income r_t for their saving. Since it is a perfectly competitive economy, the households take prices as given, i.e. they are price takers. The households spend part of their income on the composite consumption good, which is assumed to be the numeraire good and its price is normalised to 1. Consequently, this economy is akin to a pure-commodity exchange economy with the composite consumption good where money is assumed to act as veil or simply a medium of exchange in the market. Finally, households are assumed to hold an initial asset endowment, a_0, at the starting time period $t = 0$. Now letting w_t denote the real wage and r_t denote the real interest rate, the households' budget constraint at time t is given by

$$c_t + a_{t+1} = w_t + (1 + r_t)a_t. \tag{12.2}$$

The inter-temporal budget constraint (12.2) equates the expenditure for consumption c_t and the expenditure of purchase of assets that pay out in the next period a_{t+1} with the labour income w_t plus the principal and the interest of assets purchased in the previous period $(1 + r_t)a_t$.

Assets

All assets in this economy are real and there are no monetary assets. Capital assets are assumed to fully depreciate in one time period.

The assets owned by the households are assumed to be capital assets and in each period, the households decide to 'rent' their capital to the firms and earn interest or rental income. The firms use the capital in the production of goods and services, which in turn are bought by the households.

At the start of life, so to speak, say at $t = t_0$, the households start with (or are born with) assets a_0. Since the economy lasts only up to time T, we need to figure out what happens at the end of life or at period T. For simplicity, we assume the households do not leave any unspent assets after death, which implies that $a_{T+1} = 0$.

Capital and labour assumption

While the household owns capital assets, firms have the technology to produce output using capital and labour. Firms hire labour and borrow capital from the households. This is an inconsequential assumption and simplifies the model without taking away any substantive content of the argument.

With the above assumptions, the household's economic problem can now be stated as follows. Given the time path of wages and interest rates $(w_t, r_t)_{t=0}^{t=T}$ and initial assets a_0, the household wants to maximise its life-time utility function and solve the *inter-temporal optimisation problem*

$$\max_{n_t, k_t} \quad U(c_0, \ldots, c_T), \tag{12.3}$$

$$\text{subject to} \quad c_t + a_{t+1} = w_t n_t + (1 + r_t) a_t, \tag{12.4}$$

where

$$c_t \geq 0, \tag{12.5}$$

$$a_{T+1} = 0. \tag{12.6}$$

Assuming that the utility function $u(c_t)$ is guaranteed for all $c_t \geq 0$, for all times, and when c is close to zero

$$\lim_{c \to 0} \frac{du(c)}{dc} = \infty,$$

we want to derive the necessary and sufficient conditions for the optimal consumption choice. To that end, we set up a *Lagrangian function* as given by

$$\begin{aligned} \mathcal{L} &= U(c_0, \ldots, c_T) + \sum_{t=0}^{T} \lambda_t \left(w_t n_t + (1 + r_t) a_t - c_t - a_{t+1} \right) \\ &= \sum_{t=0}^{T} \beta^t u(c_t) + \sum_{t=0}^{T} \lambda_t \left(w_t n_t + (1 + r_t) a_t - c_t - a_{t+1} \right), \end{aligned} \tag{12.7}$$

where λ_t denotes the *Lagrangian multiplier* at time t. Setting the first-order conditions with respect to c_t, c_{t+1} *and* a_{t+1} to zero and simplifying yields,

$$\beta^t u'(c_t) = \lambda_t \tag{12.8}$$

$$\beta^{t+1} u'(c_{t+1}) = \lambda_{t+1} \tag{12.9}$$

$$\lambda_t = \lambda_{t+1}(1 + r_{t+1}) \tag{12.10}$$

Combining (12.8)-(12.10) results in the inter-temporal optimality condition or *the Euler equation*

$$\frac{u'(c_t)}{\beta u'(c_{t+1})} = 1 + r_{t+1}. \tag{12.11}$$

The Euler equation (12.11) is the household optimality condition that states that the marginal cost of foregoing current consumption (for future consumption) is equal to the marginal benefit of saving today as given by the rate of interest.

From the budget constraint (12.2) we get

$$c_t = w_t + (1 + r_t)a_t - a_{t+1} \tag{12.12}$$

and further iteration leads to

$$c_{t+1} = w_{t+1} + (1 + r_{t+1})a_{t+1} - a_{t+2}, \tag{12.13}$$

both of which can be substituted into the Euler equation and can be simplified as,

$$u'(w_t + (1 + r_t)a_t - a_{t+1}) = \beta(1 + r_{t+1})\left(u'\left(w_{t+1} + (1 + r_{t+1})a_{t+1} - a_{t+2}\right)\right). \tag{12.14}$$

Thus, the households' optimal consumption plan is derived by solving the Euler equation (12.14) by taking wages and interest rates $(w_t, r_t)_{t=0}^{t=T}$ as given. Equation 12.14) is a second-order difference equation in terms of households' assets, i.e. a_t, a_{t+1}, a_{t+2}, with a given initial condition a_0 and the terminal condition $a_{T+1} = 0$. Next, let us derive the optimality conditions for the firms.

The Euler equation

Equation (12.11) is known as the *Euler equation*, which provides the solution to the households' own optimisation problem. It is the households' own potential supply (of labour) and demand (for goods) decision that maximises its own utility. The Euler equation is a difference equation (or in the continuous case, a differential equation) that describes the evolution of economic variables along an optimal path. The name recognises the pioneering works of mathematicians Leonhard Euler, Joseph-Louis Lagrange and Daniel Bernoulli in the 1750s, in the context of the study of classical dynamics of physical objects.

FIRMS

Like for the households, we assume all firms are identical and normalise the total number of firms to 1. The representative firm in each period hires n_t number of labour and rents k_t amount of capital (machines) from the households and produces the consumption good y_t. The production technology is described by the standard *Cobb-Douglas production function*

$$y_t = Ak_t^{\alpha} n_t^{1-\alpha}, \quad 0 < \alpha < 1, \tag{12.15}$$

where A is the technology parameter. The parameter α measures the contribution (and hence the share of) capital in production and the share of labour in production is thus given by $1 - \alpha$.

Physical capital

Recall that the asset the households save is the physical capital, which is being rented to (or invested in) the firm.

We let the rental price per unit of capital be μ_t. Taking depreciation of physical capital into account, the effective rental income that the household receives is $r_t = \mu_t - \delta$, where δ is the depreciation rate. The representative firm takes wages w_t, the rental rate of capital is r_t, and maximises its profits at time t as given by

$$\max_{n_t, k_t} \quad y_t - w_t n_t - \mu_t k_t, \tag{12.16}$$

$$\text{subject to} \quad y_t = Ak_t^{\alpha} n_t^{1-\alpha}, \tag{12.17}$$

where

$$k_t \geq 0, \tag{12.18}$$

$$n_t \geq 0. \tag{12.19}$$

Since we can transform the constrained optimisation problem to an unconstrained problem by substituting the constraint in the objective function, the maximisation problem can be written

$$\max_{n_t, k_t} \quad Ak_t^{\alpha} n_t^{1-\alpha} - w_t n_t - \mu_t k_t, \tag{12.20}$$

subject to
$$w_t = (1-\alpha)A\left(\frac{k_t}{n_t}\right)^{\alpha}, \tag{12.21}$$

$$\mu_t = \alpha A\left(\frac{k_t}{n_t}\right)^{1-\alpha}, \tag{12.22}$$

where

$$k_t \geq 0, \tag{12.23}$$
$$n_t \geq 0. \tag{12.24}$$

The optimality conditions for the firms imply that they set their wage rate equal to the marginal product of labour and the rental rate of capital equal to the marginal product of capital.

COMPETITIVE EQUILIBRIUM

The *competitive equilibrium* is defined as the state where both households and firms maximise their objective functions, subject to their constraints, and markets clear. The markets in this economy are the labour, capital and goods markets. In a competitive equilibrium all these markets have to clear, where demand equals supply, for each time period.

The labour market clearing condition simply states that the demand for labour by the representative firm n_t equals the supply of labour by the representative household. Since we have assumed that the household supplies one unit of labour in each period, the labour market clearing condition is

$$n_t = 1. \tag{12.25}$$

In the capital rental market, the demand for capital by the representative firm at t is given by k_t and the household's asset holdings at t equals a_t. Therefore, the capital market clearing condition becomes

$$a_t = k_t. \tag{12.26}$$

Walras' law

Given the prices (w_t, r_t, μ_t, p_t), as the both labour and capital markets clear with $n_t = 1$ and $a_t = k_t$ for all t, and as the households and firms achieve their respective optimal choices, the goods market also clears for all t.

The goods market clearing condition again is simply demand equals supply. The demand for goods arises from households' expenditure on consumption goods c_t plus their expenditure (or investment) on future assets a_{t+1}. Using the household budget constraint

$$c_t + a_{t+1} = w_t n_t + (1 + r_t) a_t \tag{12.27}$$

and using $a_t = k_t$ and $a_{t+1} = k_{t+1}$ we can rewrite it as

$$c_t + k_{t+1} = w_t n_t + (1 + r_t) k_t. \tag{12.28}$$

Since $r_t = \mu_t - \delta$, we have

$$c_t + k_{t+1} - k_t = w_t n_t + (\mu_t - \delta) k_t \tag{12.29}$$

or

$$c_t + k_{t+1} - (1 - \delta) k_t = w_t n_t + \mu_t k_t \tag{12.30}$$

Starting with the representative firm's problem, substituting for w_t and μ_t, and further simplifying yields

$$c_t + k_{t+1} - (1 - \delta) k_t = w_t n_t + \mu_t k_t = A k_t^{\alpha} n_t^{1-\alpha}, \tag{12.31}$$

which is the *goods market clearing condition*.

The Euler's product exhaustion theorem

When labour and capital are rewarded exactly equal to their respective marginal products they will exhaust the total product, so that

$$w_t n_t + \mu_t k_t = y_t = A k_t^{\alpha} n_t^{1-\alpha},$$

which is referred to as the *Euler's product exhaustion theorem*.

Finally, we can characterise the equilibrium allocation with a view to assessing how well the model describes reality. We can do this by substituting the labour market, capital market and goods market clearing conditions in the household optimality condition, i.e. the Euler equation (12.14).

First, from the capital market clearing condition we have $k_t = a_t k_{t+1} = a_{t+1}$ and $k_{t+2} = a_{t+2}$. Then we know from the optimality conditions of the representative firm that

$$w_t = (1-\alpha)A\left(\frac{k_t}{n_t}\right)^{\alpha} = (1-\alpha)Ak_t^{\alpha}, \tag{12.32}$$

$$r_t = \mu_t - \delta = \alpha A\left(\frac{k_t}{n_t}\right)^{1-\alpha} - \delta, \tag{12.33}$$

and similarly for w_{t+1} and r_{t+1}. Substituting (12.32)–(12.33) in the household Euler equation (12.14) and further simplifying yields

$$\begin{aligned} &u'\left(Ak_t^{\alpha} + (1-\delta)k_t - k_{t+1}\right) = \\ &\beta\left(1+\alpha Ak_{t+1}^{\alpha-1} - \delta\right)u'\left(Ak_{t+1}^{\alpha} + (1-\delta)k_{t+1} - k_{t+2}\right), \end{aligned} \tag{12.34}$$

which yields the optimal path or the steady state solution for the intertemporal optimisation problem. Note that equation (12.34) is a second-order difference equation in capital stock, i.e. k_t, k_{t+1} and k_{t+2}, that requires the parameters α, A, δ and β, the derivative of the utility function $u(c)$ that needs to be specified, along with the initial condition for $k_0 = a_0$ and a terminal condition $k_{T+1} = a_{T+1} = 0$.

12.3 ANALYSIS AND VISUALISATION

The final version of the Euler equation (12.34) is the solution to the problem of market exchange where, at any given prices, both households and firms find their respective optimal plans and the economy is at its equilibrium. In this homogenous representative agent set up, it is also the *equilibrium path,* or the optimal path, for the economy as a whole. In an "ideal" world, given the assumptions such as agents are rational, the markets are perfectly competitive, frictionless, and a world without any uncertainty, the optimal path of the model will always yield. However, the real business cycle model provides a simple way to describe the reality of fluctuations and cycles in real economic data such as GDP, employment etc. The market clearing Euler equation (12.34) brings forth the essence of the real business cycle argument in a succinct way where the business cycle fluctuations are nothing but optimal responses of rational agents for shocks. Exogenous shocks unsettle households and firms off the optimal path and generate fluctuations in the short run. If the shocks are temporary, there will be fluctuations around the optimal path whereas permanent shocks will lead to a shift in the optimal paths. In any case, the real business cycle theory argues that the fluctuations and business cycles are fundamentally driven by exogenous shocks to various parameters in the model. The dominant candidate used in this literature is the technology shock, i.e. perturbing the term A in the production function, although other parameters like the discount factor β, preference parameter σ etc., are also used.

The final Euler equation (12.34) cannot be solved analytically. The general method used in the literature is to log-linearise the system around a 'steady state' path, where all variables are growing at the same rate. Then, using the Taylor approximation and the perturbation method, the model is used to study the impact of various shocks or impulses. The shocks (impulses) are applied to the technology and other parameters, and the resulting response of the system, in terms of both return time to the steady state and the nature of convergence to the steady state path, is analysed. In this chapter, we adopt an alternative approach of solving the inter temporal optimisation problem (12.34) using numerical methods and approximate the optimal solution to the problem. This approach seems more intuitive since the original problem is a dynamic optimisation problem, whose solution is an optimal or steady state trajectory, and business cycle fluctuations are deviations from the optimal trajectory.

In the following, we will use three parameters to introduce shocks in the model and analyse and visualise them. First, we shall find the solution for the Euler equation (12.34) using numerical methods and determine the market clearing optimal path for the economy. For this we need to assume a specific utility function $u(c)$ for the households, which is given by

$$u(c) = \frac{c^{1-\sigma}}{1-\sigma} > 0, \quad u'(c) = c^{-\sigma} > 0, \quad u''(c) = -\sigma c^{-1-\sigma}, \quad 0 < \sigma < 1, \quad (12.35)$$

where the parameter σ is the preference parameter and the perceived contribution of future utility is less than current consumption.

The Euler equation (12.34) is a second-order difference equation, with two boundary conditions, one at $t = 0$ and one at $t = T + 1$, i.e. at the beginning and the end of the life span. In economic terms, this means that the capital stock is time dependent and has memory, which can be seen in (12.34), where k_t, k_{t+1} and k_{t+2} depend on one another.

12.3.1 SOLUTION METHOD

The solution of nonlinear discrete boundary value problems like the Euler equation (12.34) described here can get quite technical, but we will provide a qualitative description of the pathway. There are a number of ways in which one can solve these kinds of boundary value problems such as *shooting* methods, *finite difference* methods and *collocation* methods. Here we will use a finite difference method, since we have a natural discretisation in k that is defined at discrete times t. This means that we need to find the entire path simultaneously for all time steps by solving a multidimensional nonlinear problem. Since the number of unknowns is relatively large and the equation is nonlinear, we cannot find a solution analytically. Instead we will use the *Newton-Raphson method* to approximate the optimal path.

First we discretise the the optimal path over $T + 1$ years in $T + 2$ time steps that are given by

$$a_0 = k_0, k_1, \ldots, k_{t-1}, k_t, k_{t+1}, \ldots, k_T, k_{T+1} = 0,$$

which means that the problem has T unknowns, since the boundary values $k_0 = a_0$, $k_{T+1} = 0$ are known. Now, letting

$$\Phi(k) = \left(\Phi_0\left(k_0\right), \Phi_1\left(k_0, k_1, k_2\right), \ldots, \Phi_T\left(k_{T-1}, k_T, k_{T+1}\right), \Phi_{T+1}\left(k_{T+1}\right)\right)^T$$

and

$$k = \left(k_0, k_1, \ldots, k_T, k_{T+1}\right)^T$$

we can rewrite (12.34) as a system of equations

$$\Phi(k) = \mathbf{0} \quad \text{or} \quad \begin{pmatrix} \Phi_0\left(k_0\right) \\ \Phi_1\left(k_0, k_1, k_2\right) \\ \vdots \\ \Phi_T\left(k_{T-1}, k_T, k_{T+1}\right) \\ \Phi_{T+1}\left(k_{T+1}\right) \end{pmatrix} = \begin{pmatrix} 0 \\ 0 \\ \vdots \\ 0 \\ 0 \end{pmatrix} \tag{12.36}$$

that we want to solve, where

$$\Phi_0 = k_0 - a_0,$$
$$\Phi_t = u'\left(Ak_{t-1}^{\alpha} + (1-\delta)k_{t-1} - k_t\right) - \beta\left(1 + \alpha A k_t^{\alpha-1} - \delta\right)u'\left(Ak_t^{\alpha} + (1-\delta)k_t - k_{t+1}\right),$$
$$\Phi_{T+1} = k_{T+1} - a_{T+1},$$

for $t = 1,...,T$.

The Newton-Raphson method for finding the optimal path k^* can now be written

$$k^{n+1} = k^n - \left(D\Phi\left(k^n\right)\right)^{-1}\Phi\left(k^n\right), \quad n = 0,1,\ldots$$

where

$$D\Phi(k) = \begin{pmatrix} \frac{\partial \Phi_0(k_0)}{\partial k_0} & \cdots & \frac{\partial \Phi_0(k_0)}{\partial k_{T+1}} \\ \vdots & \ddots & \vdots \\ \frac{\partial \Phi_{T+1}(k_{T+1})}{\partial k_0} & \cdots & \frac{\partial \Phi_{T+1}(k_{T+1})}{\partial k_{T+1}} \end{pmatrix}$$

is the Jacobian for Φ (k) that is needed for the method.

Initial conditions

To solve the optimisation problem, using the Newton-Raphson method, we need an initial guess k_t^0 for the optimal path. There are many different approaches to make an initial guess, but the problem is that we do not know anything else regarding the optimal path other than the values k_0 and k_{T+1}. Therefore, a basic approach is to discretise the initial guess linearly between k_0 and k_{T+1} so that

$$k_t^0 = a_0\left(1 - \frac{t}{T+1}\right), \quad t = 0,1,\ldots,T+1,$$

which forms a straight line between $k_0^0 = a_0$ at $t = 0$ and $k_{T+1}^0 = a_{T+1} = 0$ at $t = T + 1$.

Having found the initial conditions, the Newton-Raphson method is iterated until the solution k^n converges to the optimal path p^*, so that

$$k^n \to k^*, \quad n \to \infty.$$

However, since we do not know what the optimal path k^* looks like nor do we want to iterate the Newton-Raphson method for ever, the next best thing we can do is to compare two consecutive iterations of k^n and stop the iteration when the difference becomes smaller than some error ε, or

$$\| k^n - k^{n-1} \|_2^2 < \varepsilon$$

for some n.

We will not introduce the MATLAB® code for the method here; instead it can be found in Appendix B.

12.3.2 EQUILIBRIUM SOLUTION: THE OPTIMAL PATH

The equilibrium solution to the RBC model (12.34) is shown in Figure 12.1 with the parameters described below. It is the unique optimal solution that puts the representative agent economy on the *Pareto-efficient allocation path.* We can now analyse the stability of the optimal path for parametric variations, where we analyse the optimal path for different values of the technology parameter A, the discount factor β, the preference parameter in the utility function σ, the depreciation rate δ, and the capital income share parameter α.

Parameter values for the base case scenario

For the base case, the life span of the economy is $T = 100$, the technology parameter $A = 1$, the discount factor $\beta = 0.975$, the preference parameter $\sigma = 0.5$, the depreciation rate $\delta = 0.025$ and the capital income share parameter $\alpha = 0.4$. The initial condition for the Newton-Raphson method is taken to be $a_0 = 10$ and the terminal condition as $\varepsilon = 0.0001$.

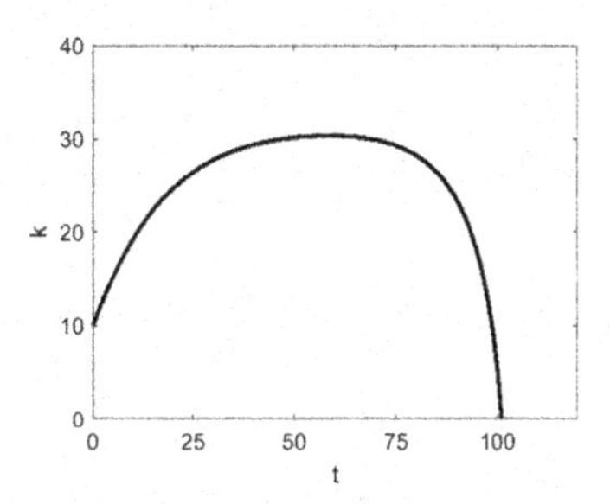

Figure 12.1: The optimal path p^* for a life cycle of $T = 100$, with $\alpha = 0.4$, $\beta = 0.975$, $\delta = 0.025$, $\sigma = 0.5$, $A = 1$ and $a_0 = 10$.

12.3.3 OPTIMAL PATHS UNDER PARAMETER VARIATIONS

From Figures 12.2 (a), (b) and (d) we can see that the optimal path is sensitive to the technology parameter A, the discount factor β and the factor income share α, respectively. On a relative scale, the optimal path is more sensitive to the discount factor β than the other two parameters. This is understandable since the discount factor influences an individual agent's allocation between consumption today and tomorrow. As the discount factor decreases, from the base case of β = 0.975, the optimal path flattens suggesting that the representative agent's consumption favours current consumption over future consumption, which is intuitive since the higher discount rate ρ puts more weight on the present vis-a-vis the future. This would in turn imply that there would be relatively less growth in capital accumulation in the lower discount factor case, for instance in the case of β = 0.9, as compared to the base case.

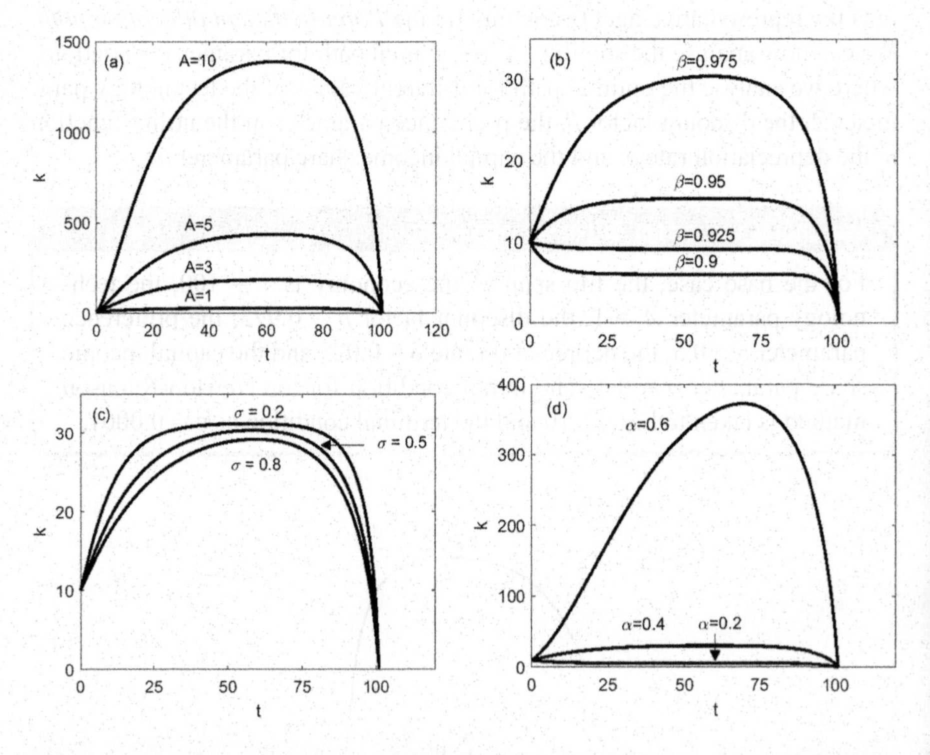

Figure 12.2: The optimal k for a life for T = 100. In (a) A is varied between 0 and 10, in (b) β is varied between 0.9 and 0.975, in (c) σ is varied between 0.2 and 0.8 and in (d) α is varied between 0.2 and 0.6.

Similarly, the optimal path is also sensitive to the technology parameter A. The technology shocks are also known as *productivity shocks* since improvements in technology, assuming these shocks last one time period, lead to an increase in the marginal product of labour and wages. This, in turn, will increase consumption and output, and the representative agent household will smooth consumption by building capital for tomorrow and thus increase the capital accumulation in the economy. This can be seen in the increase in the scale of the optimal path for higher values of the technology parameter A, as shown in Figure 12.2 (a).

The optimal path is also sensitive to α, but relatively less than for the parameters discussed above. Higher income share of capital (increasing α from 0.2 to 0.6) implies higher capital accumulation, as shown in Figure 12.2 (d). The increase in capital accumulation is via increased savings in the model. Since it is a representative agent model with the same saving rate, higher income share for capital translates to higher saving and higher capital accumulation. In a more realistic heterogeneous agent model with different saving propensities, this result may not yield as increased saving on the one hand would be countered by the adverse impact of income distribution on consumption on the other hand, and the net impact on capital accumulation may become ambiguous.

12.3.4 OPTIMAL PATHS FOR STOCHASTIC VARIATIONS IN PARAMETERS

We can also study the sensitivity of the optimal path when the system parameters are varied stochastically over the life span of the economy. Since the idea here is to provide an example of the impact of such stochastic variations in parameters on the optimal path, we choose the parameters from a uniform distribution for each time step and see how this affects the optimal path. More specifically, we perturb or shock the technology (A), agents' preference (β and σ) and other parameters (α) in the model, and assume that the shocks are drawn from a uniform distribution.

In Figure 12.3(a) the value of A at each time step is taken from the uniform distribution $U([1,1.5])$ and a single optimal path corresponding to that value is shown, while in Figure 12.3(b) for 100 different paths of A are depicted. We can see the variability, or *fluctuations,* in the optimal paths for different values of the technology parameter A. One could also extend this analysis by considering the technology shocks as having 'memory' by assuming that they follow an *autoregressive model* AR(1) process, so that $A_t = A_{t-1} + \varepsilon_t$, where ε_t can be drawn from a distribution of one's choice. Such an analysis will also yield a similar result showing the fluctuations in the optimal path.

Similarly, Figure 12.4 shows the variability in the optimal paths when the representative agent's preference parameter σ and the capital income share parameter α are shocked by drawing the respective values from different uniform distributions. In both these cases, we see that the model generates fluctuations, or business cycles. In Figure 12.5 (a) we plot the optimal paths for 100 different values of A drawn from a uniform distribution. The mean optimal path and the standard deviation of these curves are superimposed on top of the optimal curves. One can easily infer the amount of variability that the technology shocks generate in this model. This can also be seen in a more conventional way of visualizing business cycle fluctuations in Figure 12.5 (b) where we plot the standard deviation of 100,000 optimal paths for as many different values of the technology shocks drawn from the uniform distribution. We can clearly see the level of variability that the technology shocks and other parameters induce in this simple model of the real business cycle. The success of the real business cycle models is this very fact that they generate fluctuations that closely 'mimic' the real economic data. From a policy perspective, one can calibrate the model, assuming that the model is the underpinning model of the real economy, to study the potential economic cost of technology shocks and potential benefits of fine tuning the agents' preference parameters through policy interventions.

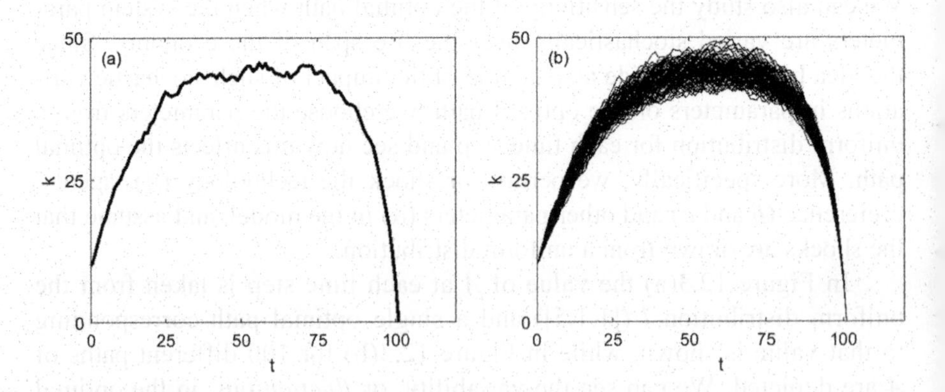

Figure 12.3: The optimal k for a life span of T = 100 years for different values of the technology parameter A taken from the uniform distribution $U([1,1.5])$. In (a) a single outcome is depicted and in (b) 100 separate optimal paths corresponding to different A values are shown.

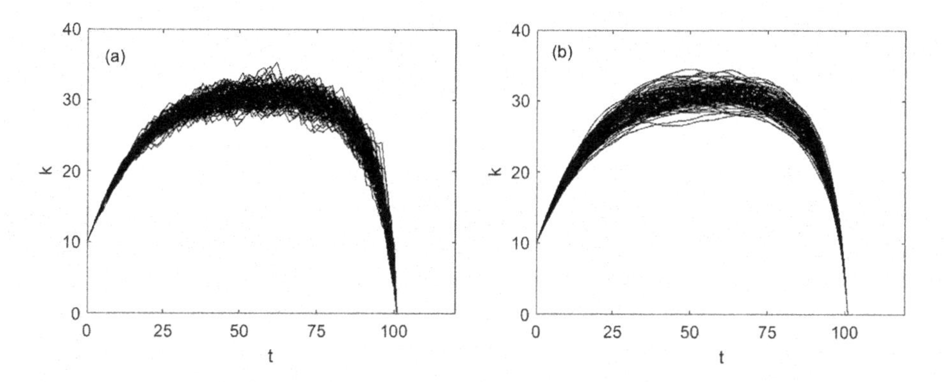

Figure 12.4: The optimal k for a life span of $T = 100$ years for 100 different σ values in (a), where σ taken from the uniform distribution $U([0.35,0.65])$, and in (b) α taken from the uniform distribution $U([0.35,0.45])$.

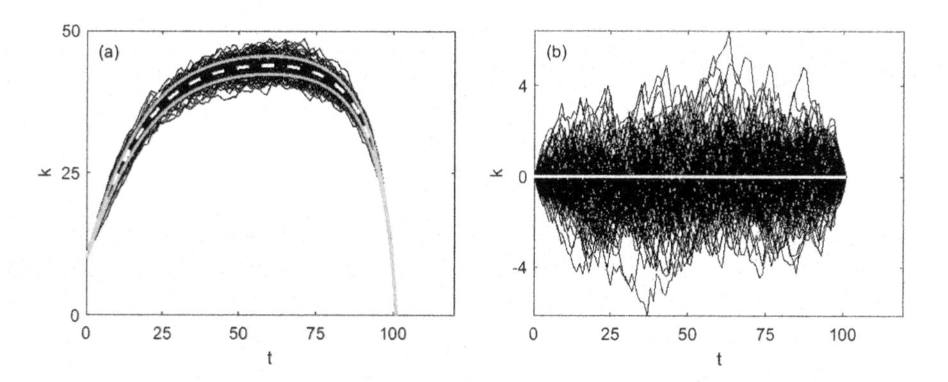

Figure 12.5: The optimal k for a life cycle of $T = 100$ years for 100000 values of A taken from the uniform distribution U ([1,1.5]). In (a) 100 optimal paths corresponding to different values of A are plotted against time. The mean (dashed curve) and standard deviation limits (solid curve) are superimposed on top of the optimal paths. In (b) the standard deviation is plotted for the 100,000 individual values of A versus time.

12.4 CONCLUDING REMARKS, EXTENSIONS AND CHALLENGES

In this chapter, we set up a basic real business cycle model and analysed the model both analytically and through numerical simulation methods. The intuition of the model is quite straightforward, namely, business cycle fluctuations are optimal responses of rational agents to unanticipated changes, or shocks. Since the representative rational agents know the underlying model of the economy and have taken all possible information into consideration in their supply and demand decisions, only unanticipated or random shocks can thwart agents' optimal plans. Therefore, business cycle fluctuations in economic data arise due to the rational agent's optimal response to unanticipated shocks. When the shocks are temporary, the fluctuations arise from consumption smoothing along the same optimal path, whereas a more permanent shock may induce shifts in the optimal path as the agent recalibrates the optimal path.

Although the model we presented here is a deterministic model, where we introduced stochastic variations in some parameters, a more general model would build stochasticity in the objective function of the households. For instance, in such a general model, at the most basic level, the households would be maximizing their *expected* life-time utility, which can be represented as

$$\max_{c_t} E\left[\sum_{t=0}^{\infty} \beta^t u(c_t)\right]$$

and firms face a production function with a time varying technology A_t, which follows an AR(1) process. With this general set-up, the competitive general equilibrium becomes a *stochastic process* and the solution of the model would entail solving the evolution of the endogenous variables and deriving the optimal path. This general model is known as the *dynamic stochastic general equilibrium* (DSGE) *model* in the literature [8,9], which has become the workhorse model for macroeconomic policy analysis. The success of the DSGE models is their ability to mimic *stylized* facts in terms of volatility, persistence and co-movements between variables that can be seen in the macroeconomic data.

However, the basic DSGE model still suffers from the shortcoming of lacking agent heterogeneity and remaining non-monetary in terms of the effects of the financial side of the economy on agents' consumption and saving decisions. The latter shortcoming became more pressing for the DSGE models post-2008 financial crisis. The newer generation of DSGE models, generically known as the *heterogeneous agent new Keynesian* (HANK) *models* [3] address some of these concerns. These models have an infinite number of households who differ in terms of labour endowments, and vary stochastically at each period, and as a consequence the capacity to accumulate savings varies over the population. The supply side is represented by a standard representative firm that determines its labor and capital demand according to the marginal return of both these production factors. The complexity in these heterogeneous agent models is such that at any given time the agents need to forecast future path of prices, which requires the knowledge of the aggregate capital stock of the economy. In a standard representative agent model, the problem can be solved in a trivial way since individual savings and aggregate capital stock coincide. However, in the heterogeneous agent set-up the aggregate capital stock depends on the distribution of savings across individual consumers. Or in other words, the distribution of wealth becomes a crucial element in the households' optimisation problem. Moreover, these complex models rely on modern computational techniques and policy implications that are sensitive to particular numerical techniques used to find solutions.[3]

In more recent times, the heterogeneous agent DSGE models are undergoing further refinements. Some of these attempts take inspiration from the complexity science and extend the DSGE models with heterogeneous agent population that can give rise to multiple equilibria resulting from the agent diversity and interaction. Professor David Vines, at the University of Oxford in the UK, issues a clarion call for such refinement in his project 'Rebuilding Macroeconomic Theory', where he proposes 'a new multiple-equilibrium and diverse (MEADE) paradigm' as the future of macroeconomics. He argues that the way forward for macroeconomics is

> . . . to start with simple models, ideally two-dimensional sketches, that explain mechanisms that can cause multiple equilibria. These mechanisms should then be incorporated into larger DSGE models in a new, multiple-equilibrium synthesis. All of this will need to be informed by closer fidelity to the data, drawing on lessons obtained from detailed work on policy models . . .
>
> (Vines and Wills [10], p. 427).

These methodological refinements provide a more secure foundation for the DSGE models. Further extensions, particularly on the substantive front in terms of integrating the monetary side of the economy and endogenising the evolution of wealth distribution in the model would help incorporate more realistic features in the DSGE models.

NOTES

1 For the purposes of this book, we will deal only with the non-overlapping finite-time horizon model. See Romer [7] for a discussion on the overlapping generations model.
2 For the purposes of description, we adapt the textbook model developed by Krueger [4]. We gratefully acknowledge Dirk for sharing the manuscript with us.
3 See Giri [2] for a survey of the heterogenous DSGE literature.

BIBLIOGRAPHY

[1] Cass, D. (1965), Optimum Growth in an Aggregative Model of Capital Accumulation, *Review of Economic Studies*, Vol. 32, pp. 233–240.

[2] Giri, F. (2017), Heterogeneity in Macroeconomics: DSGE and Agent-Based Model Approach, In Editor(s): Mauro Gallegati, Antonio Palestrini, Alberto Russo, *Introduction to Agent-Based Economics*, Academic Press, pp. 87–103.

[3] Kaplan, G., Moll, B. and Violante, G.L. (2018), Monetary Policy according to HANK, *American Economic Review*, Vol. 108, No. 3, pp. 697–743.

[4] Krueger, D. (2005), *Quantitative Macroeconomics: An Introduction*. Unpublished Manuscript.

[5] Koopmans, T.C. (1965), On the Concept of Optimal Economic Growth, In *The Economic Approach to Development Planning*, Elsevier: Amsterdam.

[6] Ramsey, F.P. (1928), A Mathematical Theory of Saving, *Economic Journal*, Vol. 38, pp. 543–559.

[7] Romer, D. (2004), *Advanced Macroeconomics, 4th edition*, Douglas Reiner.

[8] Smets, F. and Wouters, R. (2003), An Estimated Dynamic Stochastic General Equilibrium Model of the Euro Area, *Journal of the European Economic Association,* Vol. 1, No. 5, pp. 1123–1175.

[9] Smets, F. and Wouters, R. (2007), Shocks and Frictions in US Business Cycles: A Bayesian DSGE Approach, *American Economic Review*, Vol. 97, No. 3, pp. 586–606.

[10] Vines, D. and Wills, S. (2020), The Rebuilding Macroeconomic Theory Project Part II: Multiple Equilibria,Toy Models, and Policy Models in a New Macroeconomic Paradigm, *Oxford Review of Economic Policy*, Vol. 36, No. 3, pp. 427–497.

DIFFERENCE AND DIFFERENTIAL EQUATIONS

A.1 FIRST-ORDER LINEAR DIFFERENCE EQUATION

Consider a first-order linear difference equation of the form

$$x_{n+1} = Ax_n + B,$$

where A and B are constants and $n = 0,1,2,\ldots$ Since the equation is linear, the general solution is of the form

$$x_n = x_n^H + x_n^P,$$

where x_n^P is a particular solution and x_n^H satisfies the homogeneous equation

$$x_{n+1}^H = Ax_n^H.$$

Clearly

$$x_n^H = Ax_{n-1}^H = A^2 x_{n-2}^H = \ldots = A^n x_0^H.$$

To find a particular solution we try $x_n^P = x_0^P$, i.e. a fixed-point solution independent of n. Then

$$x_0^P = Ax_0^P + B \quad \Rightarrow \quad x_0^P = \frac{B}{1-A}, \quad A \neq 1.$$

Hence, we find that the general solution for $A \neq 1$ is

$$x_n = C_1 A^n + \frac{B}{1-A}, \quad A \neq 1.$$

When $A = 1$ then try $x_n^P = nx_0^P$ to find

$$(n+1)x_0^P = nx_0^P + B \Rightarrow x_0^P = B, \quad A = 1.$$

Hence we find the general solution for $A = 1$ is

$$x_n = C_1 + Bn, \quad A = 1.$$

Next we find the behaviour of the solution for large n.

For $\underline{-1 < A < 1}$: we have

$$x_n = A^n C_1 + \frac{B}{1-A} \to x^* = \frac{B}{1-A} \quad \text{as} \quad n \to \infty,$$

i.e. the sequence converges to the fixed point particular solution. For $0 < A < 1$ the sequence is monotonically convergent whereas if $-1 < A < 0$ the sequence converges oscillatory.

For $\underline{A > 1 \text{ and } A < -1}$: we have

$$x_n = C_1 A^n + \frac{B}{1-A} \to \pm\infty \quad \text{as} \quad n \to \infty,$$

i.e. the sequence diverges away from the fixed point x^* oscillatory if $A < -1$ and monotonically if $A > 1$.

For $\underline{A = -1}$: we have

$$x_n = C_1 A^n + \frac{B}{1-A} \to \pm x_0 + \frac{B}{1-A} \quad \text{as} \quad n \to \infty,$$

i.e. the sequence alternates between $x^* - x_0$ and $x^* + x_0$, where x_0 is the initial condition for $n = 0$.

For $\underline{A = 1}$: we have

$$x_n = C_1 + nB,$$

which diverges linearly for $B \neq 0$ and constant otherwise.

A.2 SECOND-ORDER LINEAR DIFFERENCE EQUATION

Here we first consider the second-order linear homogeneous difference equation

$$Ax_{n+2} + Bx_{n+1} + Cx_n = 0, \tag{A.1}$$

for constants A, B and C, we get a quadratic equation. In order to find a solution to (A.1) we first try the solution $x_n = m^n$ and thus

$$x_{n+1} = m^{n+1} \quad \text{and} \quad x_{n+2} = m^{n+2}.$$

Substituting into (A.1), and ignoring the case $m = 0$, we obtain the following *characteristic equation*

$$Am^2 + Bm + C = 0.$$

The solution to the difference equation depends on the roots, and whether the discriminant $B^2 - 4AC$ is positive, negative or zero. The general solution for (A.1) can be split into three different cases.

Case 1: $B^2 - 4AC > 0$ with real roots $(m_1 \neq m_2)$

If $|m_1| > |m_2|$ then the limiting behaviour of the sequence

$$x_n = c_1 m_1^n + c_2 m_2^n$$

is the same as that of

$$c_1 m_1^n$$

provided that $c1 \neq 0$.

Thus, if $|m_1| \geq 1$ the sequence diverges, otherwise it converges. If the sign of m_1 is negative the behaviour will be oscillatory; otherwise it will converge or diverge monotonically.

Case 2: $b^2 - 4ac = 0$ with real root m

The limiting behaviour of the sequence

$$x_n = c_1 m^n + n c_2 m^n$$

will depends on whether or not $|m| \geq 1$. If $m \geq 1$ the sequence diverges, otherwise it converges. If the sign of m is negative the behaviour will be oscillatory; otherwise it will converge or diverge monotonically.

Case 3: $b^2 - 4ac < 0$ with complex roots

In this case the solution sequence is in the form

$$x_n = r^n \left(c_1 \cos(n\theta) + c_2 \sin(n\theta)\right),$$

and the convergence of the solution depends on whether or not $|r| \geq 1$.

In all three cases we consider the roots λ_1, λ_2 of the quadratic equation

$$\lambda^2 + B\lambda + C = (\lambda - \lambda_1)(\lambda - \lambda_2) = \lambda^2 - (\lambda_1 + \lambda_2)\lambda + \lambda_1\lambda_2,$$

hence

$$\lambda_1 + \lambda_2 = -B, \ \lambda_1\lambda_2 = C.$$

The solution sequences are thus convergent in all cases provided $|\lambda_1|$, $|\lambda_2| < 1$. Hence convergence implies

$$|C| = |\lambda_1||\lambda_2| < 1.$$

We note that in case (II) $C = \lambda^2 > 0$ and in case (III) $C = \lambda_1\lambda_2 = r^2 > 0$ also.

A.3 FIRST-ORDER LINEAR DIFFERENTIAL EQUATIONS

Consider the ordinary differential equation (ODE)

$$\frac{dx}{dt} = ax \text{ or } \dot{x} = ax \quad (*)$$

with **initial condition** (IC) – the starting point for the solution –

$$x(0) = x_0.$$

For this problem we can find the explicit solution by rewriting the ODE as

$$\frac{dx}{x} = a\,dt$$

and integrating to find

$$\ln x = at + C \Rightarrow x(t) = Ce^{at}$$

for a constant C. Now we can use the IC $x_0 = x(0) = C$ to find

$$x(t) = x_0 e^{at}.$$

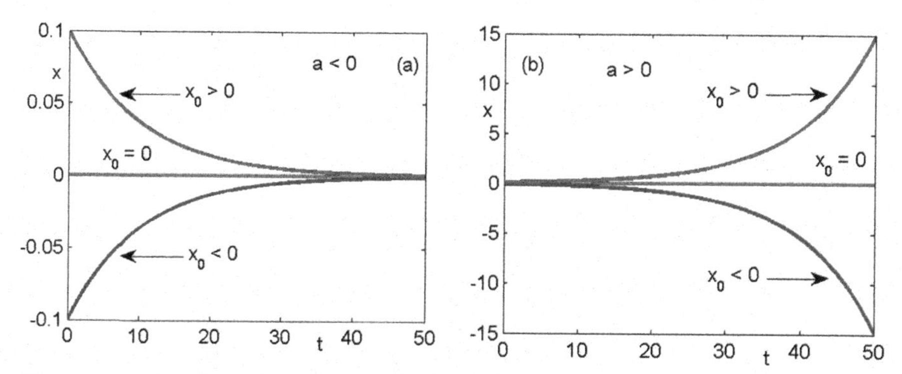

Figure A.1: (a) Solutions for $a < 0$. (b) Solutions for $a > 0$

Some useful definitions and common dynamics:

Equilibrium: A value of x (denoted by x_{eq}) for which $\dot{x} = 0$. $x_{eq} = 0$ for the above equation (*).

Equilibrium solution:

$$x(t) = x_{eq} \quad \text{for all} \quad t$$

The solution starts at $x(0) = x_0 = x_{eq} (= 0)$ and stays there for all future times, i.e. no motion occurs as times goes by.

Non-equilibrium solution: Any solution corresponding to a starting point $x_0 \neq x_{eq}$.

If $a > 0$

- $x_0 > 0$ gives $x(t) = x_0 e^{at} \to \infty$ as $t \to \infty$,
- $x_0 < 0$ gives $x(t) = x_0 e^{at} \to -\infty$ as $t \to \infty$,

i.e. all solutions move **away from** the equilibrium point.

If $a < 0$

- $x_0 > 0$ gives $x(t) = x_0 e^{at} \to 0^+$ as $t \to \infty$,
- $x_0 < 0$ gives $x(t) = x_0 e^{at} \to 0^-$ as $t \to \infty$,

i.e. all solutions move **towards** the equilibrium point.

The solutions to the ODE with IC $x_0 = -0.1, 0, 0.1$ for $a < 0$ and $a > 0$ are shown in Figure A.1

A.4 SECOND-ORDER LINEAR DIFFERENTIAL EQUATIONS

A second order linear differential equation is an equation of the form

$$a(t)\frac{d^2y(t)}{dt^2}+b(t)\frac{dy(t)}{dt}+c(t)y(t)=f(t).$$

We shall be concerned with the case where *a, b* and *c* are *constant*.

HOMOGENEOUS CASE

Let us consider the case where *a, b* and *c* are constant and where *f*(*t*) is zero so that we have

$$a\frac{d^2y(t)}{dt^2}+b\frac{dy(t)}{dt}+cy(t)=0.$$

Assume that the solution is of the form $y=e^{\lambda t}$, which leads to the characteristic equation

$$a\lambda^2+b\lambda+c=0 \tag{A.2}$$

with roots

$$\lambda_1,\lambda_2=\frac{-b\pm\sqrt{b^2-4ac}}{2a}. \tag{A.3}$$

Similarly to what we saw for first-order differential difference equations the solutions of depends on the following conditions:

(i) $b^2>4ac$, and thus $\lambda_1,\ \lambda_2\in\mathbb{R},\ \lambda_1\neq\lambda_2$
(ii) $b^2=4ac$, and thus $\lambda_1,\ \lambda_2\in\mathbb{R},\ \lambda_1=\lambda_2$.
(iii) $b^2<4ac$, and thus $\lambda_1,\ \lambda_2\in\mathbb{C},\ \lambda_1=\alpha+\mathbf{i}\beta,\ \lambda_2=\alpha-\mathbf{i}\beta$,

Now, the solution for the three cases can be written

(i) $y(t)=c_1e^{\lambda_1 t}+c_2e^{\lambda_2 t}$,

(ii) $y(t) = (c_1 + c_2 t)e^{\lambda_1 t}$,

(iii) $y(t) = c_1 e^{\alpha t}(c_1 \sin(\beta t) + c_2 \cos(\beta t)$,

where the parameters c_1 and c_2 are found from the initial conditions.

A.5 SYSTEMS OF LINEAR FIRST-ORDER DIFFERENTIAL EQUATIONS

The best way to view the system

$$\frac{dx}{dt} = ax + by$$

$$\frac{dy}{dt} = cx + dy$$

is by using matrix notation, where

$$X = \begin{pmatrix} x \\ y \end{pmatrix}, \quad \frac{dX}{dt} = \begin{pmatrix} \frac{dx}{dt} \\ \frac{dy}{dt} \end{pmatrix}, \quad A = \begin{pmatrix} a & b \\ c & d \end{pmatrix}$$

giving the matrix ODE

$$\frac{dX}{dt} = AX, \quad X(0) = X_0.$$

The properties of the matrix A determine the nature of the solutions of the equation, where the solution can be written

$$X(t) = X_0 e^{At}.$$

Notice that X_0 is a vector and e^{At} is a matrix and the solution is given by the multiplication between X_0 and e^{At}. What does the term e^{At} mean? One way to look at it is to Taylor expand e^{At}, such that

$$e^{At} = 1 + At + \frac{A^2 t^2}{2!} + \frac{A^3 t^3}{3!} + \cdots.$$

The difficulty lies in identifying the matrix representation of e^{At}. Some, but not all, 2×2 matrices can be *diagonalised* with their eigenvalues along the diagonal. This corresponds to a linear transformation of the vector space to a different set of basis vectors, aligned with the eigenvectors of the matrix. This means that the stability of the linear system is given by the eigenvalues of A.

To find the eigenvalues $\lambda_{1,2}$ we solve the **characteristic equation** of A:

$$\det(A - \lambda I) = 0,$$

where I is the identity matrix, and if expanded out

$$\lambda^2 - \mathrm{tr}(A)\lambda + \det(A) = 0 \quad \text{or} \quad \lambda^2 - \tau\lambda + \delta = 0$$

if use the notation

$$\tau = \mathrm{tr}(A) = a + d \quad \text{and} \quad \delta = \det(A) = ad - bc.$$

The solution to this equation are the eigenvalues of A, i.e.

$$\lambda_{1,2} = \frac{1}{2}\left(\tau \pm \sqrt{\tau^2 - 4\delta}\right).$$

The equilibrium is stable if the real parts of both λ_1 and λ_2 are negative and unstable if the real parts of **one** eigenvalue is positive. Here we have:

$\tau^2 > 4\delta$**:** Two distinct real eigenvalues and two distinct eigenvectors.
$\tau^2 = 4\delta$**:** Two equal real eigenvalues and two distinct eigenvectors **or** one real eigenvalue and one eigenvector.
$0 < \tau^2 < 4\delta$**:** Two complex eigenvalues and two (complex) eigenvectors.
$\tau = 0, \delta > 0$**:** Two imaginary eigenvalues and two (complex) eigenvectors.

B MATLAB® CODES

B.1 CHAPTER 1

myfirstplot.m

```
x = linspace(0,2,100);        % variable vector x with
                              % 100 elements.
y = sin(2*pi*x);              % output

plot(x,y,'k','LineWidth',2) % Plot of x versus y
axis([0 2 -2 2])              % Set figure axes

xlabel('x'), ylabel('y')      % Name of axes
title('My first plot')        % The figure title
```

mysecondplot.m

```
x = 0:4*pi/100:4*pi; % variable vector x with 100 elements,
y1 = sin(x);         % output of the first function
y2 = 2*sin(x);       % output of the second function
y3 = sin(2*x);       % output of the third function

figure(1)                        % open a new figure window
                                 % called 1
hold on                          % do not erase old plots
                                 % when plotting
plot(x,y1,'k','LineWidth',2)     % Plot of x versus y1;
                                 % color=black (k);
                                 % solid; width of the graph 2cm
```

```
plot(x,y2,'r--','LineWidth',2)  % Plot of x versus y2;
                                % color=red (r); dashed (--)
plot(x,y3,'b:','LineWidth',2)   % Plot of x versus y3;
                                % color=blue (r); dotted (:)
xlabel('x'),ylabel('y_1, y_2, y_3') % label the axes
axis([0 4*pi -2.5 2.5])         % set figure axes
box on                          % outline box on
hold off                        % erase old plots when
                                % plotting
```

B.2 CHAPTER 2

The Cobweb model

```
%The Cobweb model

%Preamble
clear all % Clears all variables
close all % Closes all figure windows

%Parameters
d0 = 4;
d1 = -1.7;
s0 = 0.5;
s1 = 1.9;

%Intial Condition
p0 = 1;

%Simulation parameters
n = 10; % Number of iterations

%Simulation
p(1:n+1)=zeros(1,n+1);
p(1) = p0;
for t = 1:n
    p(t+1) = (s1/d1)*p(t) + (s0-d0)/d1;
end

%Plotting
figure(1)
plot(0:n,p,'ko:','MarkerFaceColor', 'k')
xlabel('t'),ylabel('p_t')

figure(2),hold on,
plot([0,d0],[-d0/d1,0],'k',[s0,d0],[0,(d0-s0)/s1],'k')
plot((d1*s0-d0*s1)/(d1-s1),(s0-d0)/(d1-s1),'ko',...
      'MarkerSize',6,'MarkerFaceColor','k',...
      'MarkeredgeColor','k')
box on, hold off
ylabel('Price (p)')
xlabel('Quantity (q)')
```

```
figure(3),hold on,
D = d0 + d1*p; %Demand
S = s0 + s1*p; %Supply

plot([0,d0],[-d0/d1,0],'k',[s0,d0],[0,(d0-s0)/s1],'k')
for k=1:n
    plot([D(k)S(k)],[p(k)p(k)],'k-o',...
        'MarkerFaceColor','k')
    if mod(k,2)==0 && k>1
        plot([S(k-1),D(k)],[p(k-1),p(k)],'k--')
    elseif k>1
        plot([D(k),S(k-1)],[p(k-1),p(k)],'k--')
    end
end
box on, hold off
axis([2.1,2.5,0.8,1.1])
%axis([1.2,2.6,0.8,1.6]) %Figure 2.4(b)
ylabel('Price (p)')
xlabel('Quantity (q)')
```

B.3 CHAPTER 3

The Cobweb model

```
%Demand and supply with expectation
%Preamble
clear all % Clears all variables
close all % Closes all figure windows

%Parameters
d0 = 4;
d1 = -1.7;
s0 = 0.5;
s1 = 1.9;
c  = 0.2;

%Intial Condition
p0 = 1;

%Normal price
p_N = (s0-d0)/(d1-s1);

%Simulation parameters
n=10; % Number of iterations

%Simulation
%p(1) = p0;
%for k = 1:n
%    p(k+1) = (s0/d1)*(1-c)*p(k) + (b1*c*p_N+s0-d0)/d1;
%end

p(1:n+1)=zeros(1,n+1);
pe(1:n+1)=zeros(1,n+1);
p(1) = p0;
for k = 1:n
    pe(k+1) = p(k) + c*(p_N - p(k));
    p(k+1) = (s1/d1)*pe(k+1) + (s0-d0)/d1;
end
```

```
%Plotting
figure(1)
plot(0:n,p,'ko:','MarkerFaceColor', 'k')
xlabel('t'),ylabel('p_t')

figure(2),hold on,
plot([0,d0],[-d0/d1,0],'k',[s0,d0],[0,(d0-s0)/s1],'k')
plot((d1*s0-d0*s1)/(d1-s1),(s0-d0)/(d1-s1),'ko',...
    'MarkerSize',6,'MarkerFaceColor','k',...
    'MarkeredgeColor','k')
box on, hold off
ylabel('Price (p)')
xlabel('Quantity (q)')

figure(3),hold on,
D = d0 + d1*p; %Demand
S = s0 + s1*pe; %Supply

plot([0,d0],[-d0/d1,0],'k',[s0,d0],[0,(d0-s0)/s1],'k')
for k=2:n
plot([D(k-1)S(k)],[p(k-1)pe(k)],'k-o',...
    'MarkerFaceColor','k')
    if mod(k,2)==0 && k>1
        plot([S(k),D(k)],[pe(k),p(k)],'k--')
    elseif k>1
      plot([D(k),S(k)],[p(k),pe(k)],'k--')
    end
end
box on, hold off
axis([2.28 2.38 0.95 1.01]) %Figure 3.1(a)
%axis([2.28 2.42 0.94 1.01]) %Figure 3.1(b)
ylabel('Price (p)')
xlabel('Quantity (q)')
```

B.4 CHAPTER 4

The Macroeconomic multiplier model

```
%Preamble
clear all %Clears all variables
close all %Closes all figure windows

%Parameters
a      = 10;
b      = 0.8;
Ibar   = 20;
Gbar   = 20;

%Intial Condition
Y0 = 300;

%Simulation parameters
n = 30; %Number of iterations

%Fixed point
Yfp = (a+ Ibar + Gbar)/(1-b);

%Simulation
Y(1:n+1) = zeros(1,n+1);
Y(1) = Y0;
for m = 1:n
    Y(m+1) = b*Y(m) + a + Ibar + Gbar;
end

%Solution
%Y(1:n+1) = zeros(1,n+1);
%Y(1) = Y0;
%for m = 1:n
%    Y(m+1) = (Y0-Yfp)*b^t+Yfp;
%end

%Plotting
figure(10),hold on
plot(0:n,Y,'k--o','MarkerFaceColor','k')
plot([0,n],[Yfp,Yfp],'k--')
xlabel('t'),ylabel('Y_t')
box on
hold off
```

B.5 CHAPTER 5

The IS-LM model

```
%IS-LM model
clear all
close all
format compact

%Parameters
c1     = 0.1;
c2     = 0.1;
s      = 0.5;
a2     = 1;
k1     = 1;
k2     = 0.1;
Ibar   = 0.55;
Mbar   = 1;
% The coefficient matrix
A = [-c1*s,-c1*a2; c2*k1, -c2*k2];
b = [c1*Ibar; -c2*Mbar]; % The constant vector
%p= [c1,c2,s,a2,k1,k2,Ibar,Mbar]; %Vector of parameters

%Simulation time
t_start = 0;
t_end = 200;

%Initial conditions
Y0 = 1;
r0 = 0.01;

%Simulation
%Simulation tolerances
options = odeset('RelTol',1e-6,'AbsTol',1e-6);
[t,Z] = ode45(@(t,y)islm_system(t,y,A,b),...
        [t_start t_end],[Y0;r0],options);

%Output
Y = Z(:,1);
r = Z(:,2);
```

```
%Equilibrium point
Yr_eq = A\(-b); %same as inv(A)*b
Y_eq  = Yr_eq(1);
r_eq  = Yr_eq(2);

%Eigenvalues
eigenvalues = eig(A);
lambda_1 = eigenvalues(1)
lambda_2 = eigenvalues(2)

%Plotting
figure(1)
subplot(2,1,1)
plot(t,Y,'k','LineWidth',2)
axis([0,20,0.98,1.04])
ylabel('Y(t)')
subplot(2,1,2)
plot(t,r,'k','LineWidth',2)
axis([0,20,0,0.08])
xlabel('time (t)'),ylabel('r(t)')

Y_LM = [1,1.008];
r_LM = kl/k2*Y_LM - Mbar/k2;
Y_IS = [0.98,1.04];
r_IS = Ibar/a2-s/a2*Y_IS;

figure(2)
hold on
plot(Y,r,'k','LineWidth',2)
plot(Y_eq,r_eq,'ko','MarkerSize',6,...
    'MarkerEdgeColor','k','MarkerfaceColor','k')
plot(Y0,r0,'ko','MarkerSize',6,...
    'MarkerEdgeColor','k','MarkerfaceColor','k')
plot(Y_IS,r_IS,'k','MarkerSize',4)
plot(Y_LM,r_LM,'k','MarkerSize',4)
axis([0.98,1.04,0,0.08])
xlabel('Y'),ylabel('r')
hold off, box on
```

```
%------------------------------------------------------------
function dYdr = islm_system(t,y,A,b)

    %Variables
    Y = y(1);
    r = y(2);

    %The ODE model
    dYdr = A*[Y;r]+b;
end
```

B.6 CHAPTER 6

Debt, deficit and stabilisation

```
%debt-deficit
clear all
close all
format compact

%Parameters
b = 1:
r = 2
g = 3

p = [b,r,g]; %Vector of parameters

%Simulation time
t_start = 0;
t_end = 10;

%Initial conditions
d0 = -0.8;

%Simulation
%Simulation tolerances
options = odeset('RelTol',1e-6,'AbsTol',1e-6);
[t,d] = ode45(@(t,y) debt_deficit_system(t,y,p),[t_start t_end],...
            d0,options);

%%Output
%Y = d(:,1);
%r = d(:,2);

%Equilibrium point
d_eq = b/(g-r); %Solve dd/dt = b+(r-g)d_ed = 0

%Plotting
figure(1)
d_vec = linspace(-2,2,100);
hold on
plot(d_vec,(r-g)*d_vec+b,'k','LineWidth',2)
```

```
plot([d_vec(1),d_vec(end)],[0,0],'k--','LineWidth',2)
plot([0,0],[d_vec(1),d_vec(end)],'k--','LineWidth',2)
plot(d_eq,0,'ko','Markersize',8,'Markerfacecolor','k')
xlabel('d'),ylabel('d')
axis([-2 1-1 2])
hold off, box on

figure(2)
hold on
plot(t,d,'k','LineWidth',2)
plot([t_start,t_end],[d_eq,d_eq],'k:','LineWidth',2)
%plot([t_start,t_end],[0,0],'k', 'LineWidth' ,2)
xlabel('time (t)'),ylabel('d(t)')
axis([0 3 -3 2])
hold off, box on

%-----------------------------------------------------------------
function dd = debt_deficit_system(t,y,p)

    %Parameters
    b = p(1);
    r = p(2);
    g = p(3);

    %Variables
    d = y(1);

    %The ODE model
    dd = b +(r-g)*d;

end
```

B.7 CHAPTER 7

Cagan – linear expectation

```
%Preamble
clear all % Clears all variables
close all % Closes all figure windows

%Parameters
alpha = 0.5; %0.75; %4;
gamma = 0.5; %0.75; %0.75;
m     = 10;

%Intial Condition
p0 = 11;

%Simulation parameters
n= 10; % Number of iterations

%Simulation
p(1) = p0;
for k = 1:n
    p(k+1) = -alpha*gamma/(1-alpha*gamma)*p(k)+...
             m/(1-alpha*gamma);
end

%Fixed point
pfp = m;

%Plotting
figure(1),hold on
plot(0:n,p,'ko','MarkerFaceColor','k')
plot([0,n],[pfp,pfp],'k--','LineWidth',2)
xlabel('t'),ylabel('p_t')
box on
hold of
```

Cagan – perfect forsight

```
%Preamble
clear all %Clears all variables
close all %Closes all figure windows

%Parameters
alpha = 4;
m     = 10;

%Intial Condition
p0 = 11;

%Simulation parameters
n= 10; % Number of iterations

%Simulation
p(1) = p0;
for k = 1:n
    p(k+1) = (1+alpha)/alpha*p(k) - m/alpha;
end

%Fixed point
pfp = m;

%Plotting
figure(1),hold on
plot(0:n,p,'ko','MarkerFaceColor','k')
plot([0,n],[pfp,pfp],'k--','LineWidth',2)
xlabel('t'),ylabel('p_t')
box on
hold off
```

Cagan – stochastic

```
%Premeable
clear all % Clears all variables
%close all % Closes all figure windows

%Parameters
alpha = 4;
m     = 10;

%Intial Condition
p0 = 10;

%Simulation parameters
n= 10; % Number of iterations

%Simulation
p(1)   = p0;
for k = 1:n
    e = randn(1,1);
    p(k+1) = 1/(1+alpha)*e + m;
end

%Fixed point
pfp = m;

%Plotting
figure(100),hold on
plot(0:n,p,'ko','MarkerFaceColor','k')
plot([0,n],[pfp,pfp],'k--','LineWidth',2)
xlabel('t'),ylabel('p_t')
box on
hold off
```

B.8 CHAPTER 8

The Dornbusch overshooting model

```
%Dornbusch overshooting model
clear all
close all
format compact

%Parameters
alpha = 1;
beta  = 1;
sigma = 1;
theta = 0.5;

J = [0 1/alpha; theta*beta -theta*(beta+sigma/alpha)];

%Equilibrium
e_eq  = 2;
p_eq  = 2;

%Simulation time
t_start = 0;
t_end = 5;

%Initial conditions
e0 = 2.7321;
p0 = 1;

%Simulation
%Simulation tolerances
options = odeset('RelTol',1e-6,'AbsTol',1e-6);
[t,Y] = ode45(@(t,y)dornbusch_system(t,y,e_eq,p_eq,J),...
              [t_start t_end],[e0;p0],options);

%Output
E = Y(:,1);
P = Y(:,2);
```

```
%Eigenvalues and eigenvectors
[eigenvectors,eigenvalues] = eig(J);
mu_1 = eigenvalues(1,1);
mu_2 = eigenvalues(2,2);

ev_1 = eigenvectors(:,1);
ev_2 = eigenvectors(:,2);

%Plotting
figure(2),hold on
%subplot(2,1,1), hold on
plot(t,E,'k','LineWidth',2)
plot(t,P,'k','LineWidth',2)
plot(0,e0,'o','MarkerFaceColor','k','MarkerEdgeColor','k')
plot(0,p0,'o','MarkerFaceColor','k','MarkerEdgeColor','k')
xlabel('time (t)'),ylabel('e(t)')
hold off, box on
subplot(2,1,2), hold on
plot(t,E,'k','LineWidth',2)
plot(t,P,'k','LineWidth',2)
plot(0,e0,'o','MarkerFaceColor','k','MarkerEdgeColor','k')
plot(0,p0,'o','MarkerFaceColor','k','MarkerEdgeColor','k')
xlabel('time (t)'),ylabel('p(t)')
hold off, box on

figure(1), hold on
plot(E,P,'k','LineWidth',2)
xlabel('e(t)'),ylabel('p(t)')
k1 = ev_1(2)/ev_1(1);
k2 = ev_2(2)/ev_2(1)
plot([0 4] ,k1*([0 4]-2)+2,'k--')
plot([0 4] ,k2*([0 4]-2)+2,'k--')
plot([2 2-1/k2],[2 1],'k-')
plot(2-1/k2,1,'o','MarkerFaceColor','k',...
    'MarkerEdgeColor','k')
plot(2,2,'o','MarkerFaceColor','k','MarkerEdgeColor','k')
plot(e0,p0,'o','MarkerFaceColor','k','MarkerEdgeColor','k')
axis([0 4 0 4])
hold off, box on
```

```
%-----------------------------------------------------------
function dYdr = dornbusch_system(t,y,e_eq,p_eq,J)

    %Variables
    e = y(1);
    p = y(2);

    %The ODE model
    dYdr = J*[e-e_eq; p-p_eq];
end
```

B.9 CHAPTER 9

Graphical stability analysis

```
%Solow dynamics
clear all
close all

%Parameters
s     = 0.8;
alpha = 0.5;
n     = 0.5;
a     = 0.3;
delta = 0.4;

%Variable r
r1 = linspace(0,1,100);
r2 = [0,1];

%Plotting
r_eq=(s/(n+a+delta))^(1/(1-alpha));

figure(10), hold on
plot(r1,s*r1.^alpha,'k','LineWidth',2)
plot(r2,(n+a+delta)*r2,'k','LineWidth',2)
plot(r_eq*[1,1],s*r_eq^(alpha)*[0,1],'k--')
plot(r_eq*[0,1],s*r_eq^(alpha)*[1,1],'k--')
xlabel('r'),ylabel('y_1, y_2')
axis([0,1,0,1])
box on, hold off
```

The Solow model

```
%The Solow model
clear all
close all
format compact

%Parameters
s = 0.8:
```

```
alpha = 0.5:
n = 0.3:
a = 0.3:
delta = 0.4:

%Simulation time
t_start = 0:
t_end = 10;

%Initial conditions
r0 = 0.9;

% Simulation setup
options = odeset('RelTol',1e-6,'AbsTol',1e-6);

% Simulation
[t,R] =...
ode45(@(t,r) solow_system(t,r,s,alpha,n,a,delta),...
      [t_start t_end],r0,options);

%Plotting
r_eq=(s/(n+a+delta))^(1/(1-alpha));

figure(1)
hold on
plot(t,R(:,1),'k','Linewidth',2)
plot([t_start,t_end],r_eq*[1,1],'k--','Linewidth',1)
axis([0,10,0,1])
box on
xlabel('t'), ylabel('r(t)')

%-----------------------------------------------------------------------
function dr = solow_system(t,r,s,alpha,n,a,delta)

%The ODE model
dr = s*r^alpha-(n+a+delta)*r;
```

B.10 CHAPTER 10

The endogenous growth model

```
%Endogenous growth model
clear all
%close all
format compact

%Parameters
alpha = 0.5;
beta  = 0.2;
gamma = 2;
theta = 0.5;
n = 1;

params = [alpha, beta, gamma, theta, n];

% Jacobian
J4 = [(1-alpha)*(eq4(2)+n-2*eq4(1)),(1-alpha)*eq4(1);...
     beta*eq4(2),beta*eq4(1)+gamma*n+2*(theta-1)*eq4(2)];

% Equilibrium points
eq1  = [0,0];
eq2  = [n,0];
eq3  = [0,gamma*n/(1-theta)];
eq4  = [n*(beta+gamma)/(1-(beta+theta))+n,...
        n*(beta+gamma)/(1-(beta+theta))];

%Initial conditions
g_K0 = 10; %8+1/3+0.0001;
g_A0 = 8.44382; %7+1/3+0.0001;
y0 = [g_K0;g_A0];

%Simulation time
t_start = 0;
t_end = -1;

% Simulation setup
options = odeset('RelTol',1e-6,'AbsTol',1e-6);
```

```
%Simulation
[t,Y] = ode45(@(t,y) endogenous_system(t,y,params), . . .
                    [t_start t_end],y0,options);

%Output
g_K = Y(:,1);
g_A = Y(:,2);

%Eigenvalues and eigenvectors
[eigenvectors4,eigenvalues4] = eig(J4);
mu4_1 = eigenvalues4(1,1);
mu4_2 = eigenvalues4(2,2);

%Plotting
figure(1), hold on, box on
plot(t,g_K,'k','LineWidth',2)
plot(t,g_A,'k','LineWidth',2)
xlabel('t'), ylabel('g_K(t), g_A(t)')

figure(2), hold on, box on
plot(g_K,g_A,'k','LineWidth',2)
%plot(eq1(1),eq1(2),'o','MarkerFaceColor','k',...
      'MarkerEdgeColor','k')
%plot(eq2(1),eq2(2),'o','MarkerFaceColor','k',...
      'MarkerEdgeColor','k')
%plot(eq3(1),eq3(2),'o','MarkerFaceColor','k',...
      'MarkerEdgeColor','k')
%plot(eq4(1),eq4(2),'o','MarkerFaceColor','k',...
      'MarkerEdgeColor','k')
xlabel('g_K(t)'), ylabel('g_A(t)')

%-----------------------------------------------------
function dy = endogenous_system(t,y,params)

%Parameters
alpha = params(1);
beta  = params(2);
gamma = params(3);
theta = params(4);
n     = params(5);
```

```
%Variables
g_K = y(1);
g_A = y(2);

%The ODE model
dg_K = (1-alpha)*(g_A+n-g_K)*g_K;
dg_A = (beta*g_K+gamma*n + (theta-1)*g_A)*g_A;

dy = [dg_K;dg_A];
end
```

B.11 CHAPTER 11

Samuelson's stability diagram

```
%Preamble
clear all
close all

%Parameters
k_end = 5;

%Intial Conditions
k1 = linspace(0,k_end,100);
k2 = linspace(1,k_end,100);
k3 = [0,k_end];

b1 = 4*k1./(1+k1).^2;
b2 = 1./k2;
b3 = [1,1]

%Plotting
figure(1),hold on
plot(k1,b1,'k','LineWidth',2')
plot(k2,b2,'k','LineWidth',2')
plot(k3,b3,'k--','LineWidth',2')
xlabel('k'),ylabel('b')
box on, hold off
axis([0 5 0 1.2])
```

The Samuelson model

```
%Preamble
clear all
close all

%Parameters
b = 0.9; %0 < b < 1 %0.9, 0.5, 0.35
k = 3; %k > 0          %0.2, 1, 3
G = 10;
```

```
%Intial Conditions
Y0 = G/(1-b)-0.1;
Y1 = G/(1-b)-0.2;

%Simulation parameters
n = 8; % Number of iterations

%Simulation
Y(1) = Y0;
Y(2) = Y1;
for t = 3:n+1
    Y(t) = b*(1+k)*Y(t-1) - b*k*Y(t-2) + G;
end

%Fixed point
Yfp = G/(1-b);

%Plotting
figure(2),hold on
plot(0:n,Y,'ko','MarkerFaceColor','k','Markersize',4)
plot([0,n],[Yfp,Yfp],'k--')
xlabel('t'),ylabel('Y_t')
box on
hold off
```

B.12 CHAPTER 12

The deterministic real business cycle model

```
%Preamble
clear all

%Parameters
alpha = 0.6;
beta  = 0.975;
delta = 0.025;
sigma = 0.5;
A     = 1;

a0 = 10;

%Time steps
T = 100;
n = (0:1:T+1)';

%Varying parameters
%aa = [0.2,.4,.6]; %alpha + linspace(0,0.2,10);
%bb = [0.975,0.95,0.925, 0.9]; %beta - linspace(0,0.2,10);
%AA = [1,3,5,10]; %A + linspace(0,10,5);
%ss = [1,2,5,10];

%Imitial conditions
k0 = a0*(1 - n/(T+1)); %Linear

%Newton-Raphson
Kold = k0;

for tt = 1:1 %length(vector_name) %Choose a parameter to vary
%alpha  = aa(tt);
%beta   = bb(tt);
%sigma  = ss(tt);
%A      = AA(tt);
err     = 1;
```

```
while err > 0.0001

    G    = Gfun(Kold,a0,A,alpha,beta,delta,sigma);
    dG   = dGfun(Kold,A,alpha,beta,delta,sigma);
    Knew = Kold - dG\G;
    err  = norm(Knew - Kold,2)
    Kold = Knew;

end
figure(313), hold on
plot(n,Knew,'k','linewidth',2)

%pause
end

figure(313)
xlabel('t'),ylabel('k')

%Plot results
figure(600),hold on
plot(n,Knew,'c')
xlabel('t'),ylabel('k')

function G = Gfun(k,a0,A,alpha,beta,delta,sigma)
%Calculates the value of
%G = u'(k_t,k_t+1)-beta*u'(k_t+1,k_t+1)

k1_temp = k(1:end-2,1);
k2_temp = k(2:end-1,1);
k3_temp = k(3:end,1);

R1 = A*k1_temp.^(alpha) + (1-delta)*k1_temp-k2_temp;
R2 = A*k2_temp.^(alpha) + (1-delta)*k2_temp-k3_temp;
Q = 1+alpha*A*k2_temp.^(alpha-1)-delta;

FT0 = k(1,1)-a0;
F = R1.^(-sigma) - beta*Q.*R2.^(-sigma);
FT1 = k(end,1);

G = [FT0;F;FT1];
end
```

```
function dG = dGfun(k,A,alpha,beta,delta,sigma)
% Calculates the Jacobian of
% G = u'(k_t,k_t+1) - beta*u'(k_t+1,k_t+1)

k1_temp = k(1:end-2,1);
k2_temp = k(2:end-1,1);
k3_temp = k(3:end,1);

R1 = A*k1_temp.^(alpha) + (1-delta)*k1_temp-k2_temp;
R2 = A*k2_temp.^(alpha) + (1-delta)*k2_temp-k3_temp;

dR1_dt = alpha*A*k1_temp.^(alpha-1)+1-delta;
%dR2_dt = 0;

dR1_dt1 = -1;
dR2_dt1 = alpha*A*k2_temp.^(alpha-1)+1-delta;

%dR1_dt2 = 0;
dR2_dt2 = -1;

Q = 1+alpha*A*k2_temp.^(alpha-1)-delta;
%dQ_dt = 0;
dQ_t1 = alpha*(alpha-1)*A*k2_temp.^(alpha-2);
%dQ_t2 = 0;

dG_t = -sigma*R1.^(-sigma-1).*dR1_dt;
dG_t1 = -sigma*R1.^(-sigma-1).*dR1_dt1 -...
        beta*dQ_t1.*R2.^(-sigma) -...
        beta*Q.*(-sigma).*R2.^(-sigma-1).*dR2_dt1;
dG_t2 = -beta*Q*(-sigma).*R2.^(-sigma-1).*dR2_dt2;

N = length(k);
zM = zeros(1,N);
dG0 = zM; dG0(1,1) = 1;
dGt = spdiags([dG_t,dG_t1,dG_t2],0:2,N-2,N);
dGT1 = zM; dGT1(1,N) = 1;

dG = [dG0;dGt;dGT1];
end
```

The stochastic real business cycle model

```
close all
clear all

%Parameters
alpha0  = 0.4;
beta    = 0.975;
delta   = 0.025;
sigma0  = 0.5;
A0      = 1;

a0 = 10;

%Time steps
T = 100;
n = (0:1:T+1)';

%Imitial conditions
k0 = a0*(1 - n/(T+1)); %Linear

KK = zeros(length(n),100000);

%Newton-Raphson
Kold = k0;
for sim = 1:100000

%Pick variables randomly
%alpha = alpha0 - 0.05 + 0.1*rand(size(n));
%sigma = sigma0 - 0.15 + 0.3*rand(size(n));
A = A0 + 0.5*rand(size(n));

while err > 0.0001

    G = Gfun(Kold,a0,A,alpha,beta,delta,sigma);
    dG = dGfun(Kold,A,alpha,beta,delta,sigma);
    Knew = Kold - dG\G;
    err = norm(Knew - Kold,2);
    Kold = Knew;

end
```

```
KK(:,sim) = Knew;

%Plot results
%figure(1),hold on
%plot(n,KK,'k','linewidth',0.5)
%xlabel('t'),ylabel('k')

Kstd = std(KK,1,2);

figure(2), hold on
plot(n,Kstd','k')
xlabel('t'),ylabel('Std')

figure(3), hold on
mKK = mean(KK,2);
plot(n,KK(:,1:1000:end)-mKK(:,1:1000:end),'k')
xlabel('t'),ylabel('k')

figure(3)
plot(n,mean(KK,2),'w--')
plot(n,-Kstd,'w',n,+Kstd,'w')

figure(4),hold on
plot(n,mean(KK,2),'w--')
plot(n,mean(KK,2)-Kstd,'w',n,mean(KK,2)+Kstd,'w')
end

function G = Gfun(k,a0,A,alpha,beta,delta,sigma)
% Calculates the value of
% G = u'(k_t,k_t+1) - beta*u'(k_t+1,k_t+1)

k1_temp = k(1:end-2,1);
k2_temp = k(2:end-1,1);
k3_temp = k(3:end,1);
A1 = A(1:end-2,1);
A2 = A(2:end-1,1);
sigma1 = sigma(1:end-2,1);
sigma2 = sigma(2:end-1,1);
alpha1 = alpha(1:end-2,1);
alpha2 = alpha(2:end-1,1);
```

```
R1 = A1.*k1_temp.^(alpha1) + (1-delta)*k1_temp-k2_temp;
R2 = A2.*k2_temp.^(alpha1) + (1-delta)*k2_temp-k3_temp;
Q = 1 + alpha2.*A2.*k2_temp.^(alpha2-1)-delta;

FT0 = k(1, 1)-a0;
F = R1.^(-sigma1)-beta.*Q.*R2.^(-sigma2);
FT1 = k(end, 1);

G = [FT0; F; FT1];
end

function dG = dGfun(k,A,alpha,beta,delta,sigma)
%Calculates the Jacobian of
%G = u'(k_t,k_t+1)-beta*u'(k_t+1,k_t+1)

k1_temp = k(1:end-2,1);
k2_temp = k(2:end-1,1);
k3_temp = k(3:end,1);
A1 = A(1:end-2,1);
A2 = A(2:end-1,1);
sigma1 = sigma(1:end-2,1);
sigma2 = sigma(2:end-1,1);
alpha1 = alpha(1:end-2,1);
alpha2 = alpha(2:end-1,1);

R1 = A1.*k1_temp.^(alpha1) + (1-delta)*k1_temp-k2_temp;
R2 = A2.*k2_temp.^(alpha2) + (1-delta)*k2_temp-k3_temp;

dR1_dt = alpha1.*A1.*k1_temp.^(alpha1-1)+1-delta;
%dR2_dt = 0;

dR1_dt1 = -1;
dR2_dt1 = alpha2.*A2.*k2_temp.^(alpha2-1)+1-delta;

%dR1_dt2 = 0;
dR2_dt2 = -1;

Q = 1+alpha2.*A2.*k2_temp.^(alpha2-1)-delta;
%dQ_dt = 0;
```

```
dQ_t1 = alpha2.*(alpha2-1).*A2.*k2_temp.^(alpha2-2);
%dQ_t2 = 0;

dG_t = -sigma1.*R1.^(-sigma1-1).*dR1_dt;
dG_t1 = -sigma1.*R1.^(-sigma1-1).*dR1_dt1 -...
        beta*dQ_t1.*R2.^(-sigma2) -...
        beta*Q.*(-sigma2).*R2.^(-sigma2-1).*dR2_dt1;
dG_t2 = -beta.*Q.*(-sigma2).*R2.^(-sigma2-1).*dR2_dt2;

N = length(k);
zM = zeros(1,N);
dG0 = zM; dG0(1,1) = 1;
dGt = spdiags([dG_t,dG_t1,dG_t2],0:2,N-2,N);
dGT1 = zM; dGT1(1,N) = 1;

dG = [dG0;dGt; dGT1];
end
```

INDEX